The Science of the X-Files

THE SCIENCE OF THE X-FILES

Michael White

Michael White

Published by Legend Books in 1996

An imprint of Random House UK Limited
20 Vauxhall Bridge Road, London SW1V 2SA

An imprint of Random House UK Limited

Random House Australia (Pty) Limited
16 Dalmore Drive, Scoresby, Victoria, 3179

Random House New Zealand Limited
18 Poland Road, Glenfield
Auckland 10, New Zealand

Random House South Africa (Pty) Limited
PO Box 2263, Rosebank 2121, South Africa

Random House UK Limited Reg. No. 954009

ISBN 0099185725

Typeset by Deltatype Ltd, Birkenhead, Merseyside
Printed and bound by Mackays of Chatham PLC

For my father, Samuel Leonard White (1915–1985),
who kindled the flame of curiosity

listen, there's a hell of a good universe next door, let's go

e e cummings

Acknowledgements

Many people have helped me during the writing of this book, but I would like to thank especially: Jaimie Tarrell, Paul Bailey and John Gribbin for their advice; my agents, Bill Hamilton and Sara Fisher for having the business brains I entirely lack, and John Jarrold, who is probably the most hospitable editor I've ever worked with. Finally, I would like to thank my wife, Lisa, who works harder than I do and kept me at the word-processor when I would have preferred to slope off to watch TV.

Preface

This is a book of hypotheses – the hypotheses upon which the *X-Files* is constructed.

There was a time, until perhaps the beginning of this century, when the investigation of the paranormal commanded a certain respect and was conducted by professional scientists as much as by enthusiastic amateurs. Such open-mindedness ended for two reasons. Firstly, many paranormal phenomena seemed so elusive that busy physicists, chemists and biologists tired of trying to pin them down. Secondly, science became so overwhelmingly successful that, to many, the supernatural became almost superfluous. Nuclear physics, brain surgery and the advancement of space travel can be as exciting as hunting ghosts or trying to prove the existence of alien civilisations, and at the same time they are tangible with commercial and academic application.

But the paranormal remains elusive. We are no nearer proving or disproving the existence of telepathy, psychokinesis or clairvoyance than we were over one hundred years ago, but we do have a greater collection of scientific ideas to draw upon in an effort to reach sensible hypotheses. And that is what I've tried to do in this book.

I have always been interested in the paranormal. I went through a period of intense enthusiasm as a youth and then into the deepest, most cynical scepticism. Now, I like to think I'm beginning to find a balanced viewpoint, the famed open-mindedness.

Like many people I have had the good fortune to experience the paranormal first-hand and, in keeping with almost everyone else in this position, I have no real answers that satisfy or explain completely what happened to me.

In 1974, I was a scientifically-minded, swotty fifteen-year-old at a traditional English private school. During the summer holiday, some of my friends, who shall remain anonymous to preserve their own embarrassment, began to experiment with a Ouija board. They made their own out of bits of paper and an orange-juice tumbler, but where their experience really differed was not in approach but what happened next. Instead of contacting Mozart or Julius Caesar, they immediately began a conversation with something that claimed to be an alien who lived on Saturn (in a parallel universe, in which the planet was habitable), and called himself Alan Kalak 7. He told the boys that he was the leader of an alien committee who were gathering together twelve young people on Earth to form *The Group* – a collection of people who would one day radically alter the future of the planet.

Now, this was pretty heady stuff for a bunch of immature boys living through acne trauma, cricket fixation and looming O-levels. It was even more sensational for me when the original contactees casually informed me when we all returned to school that Alan had included my name as a member of *The Group*.

Sitting at the Ouija board was a most peculiar experience, at once terrifying and exhilarating. The peer pressure not to make a complete prat of oneself was tremendous. But now, over two decades later, only one sitting sticks out in my memory and is the reason for this tale.

One freezing afternoon (shortly after mock O-level maths, I seem to recall), one of my friends (and the originator of the adventure), someone I'll call G, invited me to a session of the board in which just the two of us would attempt to talk to Alan.

This had never been done by any of us before. Usually there were three or four of the group and so we were always able to convince ourselves that someone else was moving the glass and that it was all really a daft game.

G set out the bits of paper on a glass table he had somehow commandeered for the purpose. He put out a simple YES, a NO and a few letters and numbers. Then we sat on opposite sides of the table and placed our fingers very lightly on the top of the glass.

Suddenly, the room seemed very quiet, sounds from outside died, almost as though they had been on a tape that had just been switched off. We could hear each other breathing heavily. Then G said: 'Alan are you there?'

For a moment nothing happened. The seconds ticked by – I could hear them clanging away at my wrist. Apart from that, silence, as we held our breath.

This was ridiculous, I began to think. Obviously, nothing was going to happen, unless of course . . . unless G had been pushing the glass all this time. But, just as I was beginning to think we were wasting our precious revision time and that I ought to call a halt to the proceedings, the glass moved.

At first, the movement was almost imperceptible, then it began to stagger and then to glide almost effortlessly across the smooth table top.

I was struck speechless. I had seen this before, but only with a group of us around the table. I looked at G's finger and it was almost hovering above the glass, hardly touching it. I looked at my own finger. Was I pushing it without realizing it? No, of course I wasn't. I too was hardly making contact with the glass.

G asked Alan a few questions – something about *The Group* and the people involved, and the glass spelt out initials, answered YES or NO as appropriate, but I was not really concentrating: I was too stunned.

Afterwards, G and I sat and talked about what had happened and if I needed any further proof that he was as innocent of fraud as I, it came from that conversation – he was as shocked and as exhilarated as me.

I would like to report that great things came from *The Group*, that there was huge significance in the contact, that Alan Kalak 7 and his chums really were a committee of aliens from a parallel universe, but unfortunately I cannot. We all went our separate ways after we finished school, and gradually, degrees, women and careers loomed larger than supposed alien contact. There were adventures and further mysterious happenings during what remained of our schooldays, but no cigar, and if anything, the planet is in a worse state now than it was in 1974. But, crucially, the memory of that afternoon remains and has nurtured enthusiasm even through my most empirical and sceptical moments.

I still have no idea what happened in 1974. I am not prepared to believe that we were contacted by aliens, or mischievous spirits, neither am I willing to consider that our untrained minds were capable of psychokinesis. Equally, to simply conclude that things happened via forces forever beyond our understanding strikes me as an unacceptable cop-out.

The fact is, I don't know how this and other incidents happened, but, because I don't know, I continue to be curious, determined that one day, I will.

Michael White, London, July 1996

Chapter 1: Visitors

> Any sufficiently advanced technology is indistinguishable from magic.
>
> Arthur C. Clarke

One of the central themes of the *X-Files* is that our planet is being visited by aliens and that some of these aliens abduct humans on a regular basis.

This is not simply a product of the programme-makers' imagination, but an idea that has become so entrenched in a range of Earth cultures as to have become almost a cliché. Whole forests have gone under the axe to help produce countless books and magazine articles on the subject and an entire mythology has been created covering every aspect of alien visitation. The only thing so far missing is hard, irrefutable proof to support the idea.

Beyond the *X-Files* itself, the most active site for information concerning alien visitors is the Internet. If you want the latest on the subject, try a newsgroup called alt.alien.visitors on the World Wide Web. Here you will encounter telepaths from the Pleiades, Reptoids from Sirius, beautiful Venusians and a ubiquitous group commonly called the greys.

The level of material now circulating about alien visitors is staggering and has become a thriving cottage industry in itself. Magazines available at newsagents in every major city in the western world have progressed far beyond rather tame interviews with abductees to giving their readers detailed descriptions of alien propulsion systems and intricate anatomical studies of alien physiology.

The source of much of this material is the Roswell incident. According to enthusiasts, a UFO crashed in Roswell, New Mexico in 1947 and believers claim the investigation into the crash has been deliberately kept from the public ever since. They also insist that US government agencies have conducted experiments on both the craft and the dead aliens found in the wreckage at a top secret location called Area 51 in the Nevada Desert and that information leaked from there substantiates their claims.

There are a range of supposed origins for alien visitors. Enthusiasts who have become disillusioned with the possibilities of interstellar travel suggest that flying saucers come either from inside the Earth itself or from 'other dimensions'. The Hollow Earth cult has been in existence for some time, but remains little more than a fantasy. The notion of 'other dimensions' is more intriguing but no less ambiguous. What are these dimensions?

It is often the case that enthusiasts of the occult use expressions like this with little or no understanding of what they mean. They then compound the problem when they try to relate such vague ideas to legitimate science, claiming that, because physicists talk about universes comprising of 10 or 26 dimensions, that this somehow accounts for their misguided theories. It doesn't – the extra dimensions currently fashionable with physicists crop up in an exotic area of physics called string theory. They are not 'parallel universes', but are thought to exist only with the sub-atomic scales discussed in the field of quantum mechanics, (sizes in the region of 10^{-33} cm). It is extremely unlikely that they could provide a possible location for alien intelligence arriving here in physical craft.

Consequently, in this chapter, I will restrict my discussion to the notion that aliens may be coming here from other planets beyond our own solar system and how they could possibly do this operating within the known laws of physics.

The facts of interstellar travel all revolve around distance, time and power. Because the distances between the stars are unimaginably huge, the time needed to travel interstellar distances is correspondingly large, and any system which may have a chance of overcoming this restriction requires impractical amounts of power.

The problem begins with Einstein's special theory of relativity. First published in 1905, when Einstein was working in a Bern patent office, the special theory draws upon two firmly established scientific principles but comes up with one of the weirdest notions in the whole of science.

The first of these derives from the work of Isaac Newton, who, back in the 1680s, showed that the laws of physics are the same for any observers moving at a constant velocity relative to one another. The second fact, arrived at more recently, is that the speed of light in a vacuum is always constant. This velocity is represented by the symbol c and is equal to just over one billion kilometres per hour. This is true, *irrespective* of the velocity of the observer.

According to common sense, if spaceship A is moving in one direction with a velocity of 0.75c and spaceship B approaches in the opposite direction also travelling at 0.75c, their relative velocity would be 1.5c. But this is not actually the case. According to Einstein's equations, crews on each ship would see light from the other coming towards them, not at one and a-half times the speed of light, but just under 1c (0.96c to be precise).

The astonishing consequence of this is that if c is constant, space and time must be relative. In other words, if the crew aboard spaceship A or B are to see light arriving at a constant velocity irrespective of their own velocity, they must measure time differently – so, as they travel faster, time slows. Furthermore, the property of distance cannot be the same to observers travelling at different speeds. The faster one travels, the shorter any given distance becomes – a metre will be a different length depending on the velocity of the observer, and will be shorter the faster the observer moves. Finally, the faster an observer moves, the more massive they become. The end result of all this is that if it were possible for an observer to travel at the speed of light they would experience three things – time would slow to nothing, they would shrink to nothing and their mass would be infinite!

Sadly for space-travel enthusiasts, this is not the delusion of a mad professor. Einstein's special theory of relativity has been proven to be true in many thousands of experiments conducted since 1905. The reason we do not notice this effect every day of our lives is that we do not travel anywhere nearly as fast enough to notice. A recent shuttle mission showed how minuscule the effect is at low speeds. Travelling in orbit at a sprightly five miles per second, clocks aboard the shuttle ticked less than one ten-millionth of a second slower than their counterparts on Earth. At CERN, the giant particle accelerator near Geneva in Switzerland, subatomic particles are accelerated to near-light speeds routinely and their masses seen to increase precisely as Einstein's calculations predict.

So, the law that states that no material object can travel at the speed of light is irrefutable; it is a fact of life in our universe. Consequently, the only possible ways a technologically advanced civilization could cross interstellar distances is either to travel at speeds which do not incur too many problems from Einstein's theory, but get them there eventually, or else they would have to find ways around the theory.

First, let's look at the sub-light speed options.

Since Jules Verne's idea of firing a moon rocket from a 900-foot deep hole in Florida in *From the Earth to the Moon* (published in 1865), scientists and science-fiction writers have come up with a range of ingenious propulsion systems to facilitate interstellar travel. These include fusion drives, antimatter engines, spaceships utilising the properties of wormholes and space-warping devices.

All conventional space-propulsion systems (and by that I mean engines that do not use some exotic property of space itself such as warping or wormholes) must work on the principle of Newton's third law of motion, which states that: 'For every action there is an equal and opposite reaction.' In this way a spaceship is no different to a jet aircraft – material is expelled from the back of the craft and the craft moves forward; simple. The difficulty is a question of magnitude.

The spacecraft we have developed so far all work by chemical propulsion. The greatest energy requirement has so far been that needed to escape the Earth's gravitational pull – to achieve an escape velocity so that the Saturn V, the shuttle or the Ariane craft can get their payloads into orbit. All manoeuvres aboard the Apollo craft travelling to the moon depended upon relatively small engines and thrusters that expelled hot gases from their exhausts and adjusted the course of the spaceship. Without these the capsules would have been entirely at the whim of the gravitational forces at work beyond the Earth's atmosphere.

The next level of sophistication is some form of fission-powered spacecraft engine. This is the power source used in nuclear reactors and unleashed in the earliest atomic bombs. When large unstable atomic nuclei are made to decay, or undergo fission, they produce energy. The value of this energy depends upon the mass of material undergoing fission and can be calculated using perhaps the most famous equation in history: $\mathbf{E = mc^2}$, where **m** equals the mass of material and **c** is the speed of light.

Although this is the most powerful controllable energy source we have currently, it could not provide anything like the energy needed to reach the stars and the mass of fissionable material needed even for efficient interplanetary travel within our own tiny Solar System would be so large, there would be little room left for crew or cargo.

A more powerful form of nuclear energy comes from a process called nuclear fusion. Back in 1989 there was a brief flurry of excitement when two scientists, Martin Fleischmann and Stanley Pons claimed they had devised a technique called 'cold fusion' which appeared to require nothing more than a pair of electrodes and some commonplace chemicals placed in a jar. Sadly, the excitement died when the experiments proved unrepeatable and the hopes of scientists returned to conventional fusion. This is a mechanism by which the Sun or any star is powered. In the laboratory the process involves fusing together small nuclei such as deuterium and tritium (which are heavy isotopes of hydrogen) to produce large amounts of energy.*

For almost fifty years scientists have been trying to develop practical nuclear fusion – it is relatively clean because it does not use dangerously radioactive elements such as the uranium-238 which is converted into plutonium-239 in modern fast breeder reactors, (isotopes that remain dangerous for hundreds of thousands of years) and it could potentially produce far more energy than fission. These are the plus points of the system; the down side has so far been the problem of containment and efficiency. In order to bring about fusion, temperatures of around 10 million degrees are needed (the sort of temperatures produced at the Sun's core) so that the positively charged nuclei can be forced to overcome their electrostatic repulsion. This fused material exists as a super-heated plasma which cannot be kept in any form of physical container. Furthermore, the energy needed to bring about fusion has so far been much greater than the energy return, which means the system currently shows negative efficiency.

Having said that, scientists hope to crack these problems in the

* A heavy isotope is a version of an atom that has more than its usual complement of neutrons in its nucleus. The most common form of hydrogen has just one proton in its nucleus and no neutrons. The first heavy isotope of hydrogen, deuterium, has one proton and one neutron. The heaviest, tritium, has one proton and two neutrons.

Figure 1

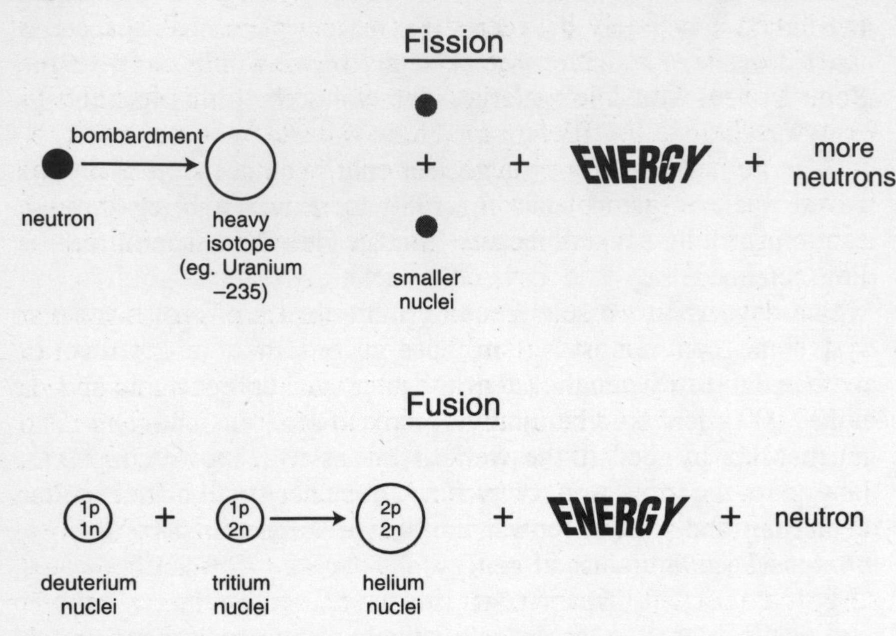

future and fusion energy is seen as the most likely way in which we could save the Earth's looming resource crisis. Assuming another civilization is only a few decades ahead of us, they would almost certainly have developed fusion power and if they are further advanced, they would have mastered the use of fusion engines aboard spacecraft. Unfortunately, this energy source could never be used for interstellar travel for the simple reason that the amount of fusible material needed to achieve even a tiny percentage of light speed would be too great.

It has been calculated that to accelerate a spaceship to just 10% of the speed of light would require about 15 times its mass in fuel. And this is to accelerate just once. If the craft wanted to stop at its destination it would need to use more fuel, equivalent to 15 times the current mass of the ship. If we assume the outward voyage has used up half the fuel (which weighed 15 times the mass of the living quarters and cargo remember) a further 7.5 times the mass would be needed. So, one start and one stop would need 15 times 7.5 times the mass of the main body of the ship (excluding fuel), or 112.5 times the mass of the living quarters and cargo.

A variation on this is the idea of the fusion ramjet. Interstellar space is not a complete vacuum, it contains hydrogen atoms, albeit distributed very finely between the stars and planets. A spacecraft could be designed with large scoops that draw in the hydrogen atoms to use as fusible material. The objection to this has always been that there is insufficient material available in space, but if the craft is moving quickly enough it would behave like a giant sea mammal drawing in plankton, or like a person running through light rain getting soaked because they are meeting the raindrops as they go.

One day, when we send people to the planets of our own Solar System, we will almost certainly use fusion power in one form or another. It is a practical system for interplanetary travel as speeds of 100,000 kph would be relatively easy to generate, allowing us to get to Mars in about three weeks. But, as we saw in Chapter 1, there is no comparison between interplanetary and interstellar distances. Using fusion power to achieve speeds of 100,000 kph, we would need a thousand generations merely to reach the nearest star, and the fuel requirements to maintain even this relatively trivial velocity for so long would alone make it totally impractical.

Putting aside fusion power, there have been a number of other suggestions for ways to achieve a reasonable fraction of light speed using conventional physics. One such idea is to use the power of nuclear explosions to thrust the craft forward.

Designers of a theoretical vehicle known as Orion visualize using a store of thermonuclear warheads individually propelled from the back of the craft at the rate of one every three seconds. The hot plasma produced by the explosions would impact on a 'pusher plate' propelling the spaceship forward. Unfortunately, to achieve a speed of just 3% of the speed of light would need almost 300,000 one-ton bombs.

A variant on this was Project Daedalus investigated by the British Interplanetary Society during the 1970s. This theoretical system involved a craft similar to Orion but powered by 250 nuclear explosions per second which could achieve some 12% of light speed, or 130 million kph, but again the mass of fuel required made the idea quite impractical.

More promising than any of these schemes could be the possibility of using exotic material known as antimatter.

All matter in our universe is made of atoms. These in turn are composed of what are called sub-atomic particles – neutrons and protons, which exist together in the nucleus of the atom, and electrons which surround the nucleus. This much was understood early this century thanks to the work of such pioneers as Ernest Rutherford, James Chadwick, Max Planck and others. Another ground-breaking physicist of this era was Paul Dirac, who in 1929 predicted that all the known subatomic particles could have counterparts with opposite properties.* These became known as antiparticles.

Protons are positively charged and an antiproton would have the same mass and exist in the nuclei of antiatoms but would be negatively charged. An antielectron, or positron as it has become known would be positively charged and like the electron exist outside the nucleus of antiatoms. But what is most important for our purposes in designing an interstellar engine is the fact that when matter and antimatter come into contact they annihilate each other instantly and produce energy.

In Paul Dirac's day, antimatter was merely a theoretical concept, something that had popped out of the equations when he had combined the mathematics of quantum mechanics, electromagnetism and relativity. At that time, the existence of antimatter could not be proven as it is not found naturally in our universe because it would disappear as soon as it came into contact with matter. Today, we can manufacture small quantities of antimatter in a particle accelerator.

To make an antiproton, 'normal' protons are sent whirling around the accelerator ring, where they are accelerated in an intense magnetic field until they reach about half the speed of light. They are then allowed to collide with the nuclei of metal atoms. This produces pairs of particles and antiparticles along with X-rays and various forms of energy. The antiprotons are then separated from the protons before they can interact and obliterate one another.

To use antimatter as a propellant we need to allow a controlled annihilation of particles and antiparticles and to use the heat evolved to drive our spacecraft. A simple design for just such a

* In those days only protons and electrons were known; the neutron was discovered three years later in 1932.

system is already on the drawing board. The idea is to fire a tiny quantity of antimatter into a hollow tungsten block which is filled with hydrogen. The particles are instantly annihilated and the energy released heats up the tungsten block. Cold hydrogen is then squirted into the centre of the device where it is rapidly heated to about 3000K and fired out of the engine.

The great advantage of antimatter drives is that little fuel is needed to produce an effective acceleration. The great disadvantage is the difficulties of producing usable amounts of the stuff. Currently only a sixty millionth of the energy used in producing antimatter in the world's particle accelerators ends up as particles, which is one of the reasons its current market value is about $10,000,000,000,000,000 (ten thousand million million dollars) per gram.

We also have the problem of containment. Like the superhot plasma produced by nuclear fusion, special magnetic containment systems have to be used, in this case to prevent antimatter interacting with matter before it is needed.

None of these difficulties precludes its use by advanced civilizations. Looking at our own history can provide a salutary lesson. It was only in 1919 that Ernest Rutherford discovered that the nuclei of certain atoms could be made to disintegrate by bombardment. Within just 26 years this discovery lead to Hiroshima and Nagasaki.

Current technology may mean that antimatter is prohibitively expensive to produce, but within two or three decades this will no longer be the case. And, such time spans are relatively meaningless when we look at the possibility of advanced societies developing on other planets.

The option of antimatter propulsion systems offers hope that interstellar travel may be a possibility, but even if an advanced civilization had realized the full potential of this technology, they would not be able to circumvent the natural laws of the universe and would remain limited to sub-light speeds.

What this means for interstellar voyagers is that they will either have to accommodate the consequences of travelling at a fraction of the speed of light (which will still be very slow for the purposes of colonizing or visiting many worlds), or else travel even slower and take even longer to arrive anywhere outside their own Solar Systems.

If we imagine for a moment a journey of 50 light years from the alien's home world to Earth. At 0.95 c (95% the speed of light), this will take 47.5 years to complete, one way. 47.5 years to the people back home, that is. Because of the consequence of special relativity that, as we travel faster, relative time slows, to the crew of the spaceship, this 47.5 years will only be 14.8 years.

This is still far too long to be of practical use. Even if we assume alien longevity is greater than ours, one and a-half decades is still a long time to be on board a spacecraft. The answer to this might be a form of suspended animation or even cryogenics, but there are other hurdles. Crews sent out on round trips of almost a century might return to their home worlds to find the political structure changed. The organization that sent them may no longer exist. Such a crew would find all their relatives either dead or ancient, and almost everything once familiar, irreversibly altered. Imagine a human able to set out on such a mission in the year 1900 returning to Earth in 1997. They may have aged less than thirty years, but the world would be almost unrecognizable to them.

If we consider, as UFO enthusiasts do, that interstellar travel is commonplace and that aliens visiting us operate as part of an organized federation or planetary authority, they could not travel at sub-light speeds. Firstly, any form of command structure would be impossible to maintain over such distances and time-scales. Secondly, it is surely a universal rule that any endeavour must see a return for the investor within a reasonable time-frame, certainly the lifetime of the investor. Who would finance any missions involving such time-scales? Certainly no government we could imagine.

The only other possible option for sub-light travel is the concept of the Ark. This has been a favourite of science-fiction writers and space analysts for generations and offers one system for interstellar travel that has some practical aspects.

The idea would be to create a large spacecraft on which generations of aliens would live during a mission that may last hundreds or thousands of years. Although this would require a vast spacecraft capable of sustaining a large number of crew and passengers for perhaps millennia, it would not need to travel particularly fast. If a mission was designed to take 1000 years, a distance of 50 light years could be covered at just 5% of light speed (50 million kilometres per hour). But, putting aside the technological difficulties of designing and constructing such a craft, there are other more mundane drawbacks with the scheme.

First, we return to the problem of time-scales. The only possible reason a civilization would finance such a mission would be to escape catastrophe – an Ark in the biblical sense, where some or all of the population of the planet are packed into spacecraft which then head off to find a new home. We have not yet encountered a colonizing group, so it would be safe to assume that our widely-reported visitors do not fall into this category. Yet, even assuming a smaller scale operation was financed and set in motion with, say, a few thousand passengers and crew, how would the group sustain itself psychologically?

The early generations of aliens aboard the Ark would have no hope of seeing a new world and would be kept going only by the knowledge that their distant offspring will make it to a distant planet. It is conceivable that such a scheme would be favoured by aliens who possess a very different psychological make-up to us. They might think in a similar way to ants or bees and have an inbuilt instinct for the community rather than the individual. Such a scheme may work for them, but places upon it severe limitations in a general sense.

But, the most persuasive argument against the use of Arks is the notion of the 'speed exponential curve'. Using our own technological development as a paradigm, it can be seen that the speeds we are capable of achieving have risen exponentially with time. For the first 100,000 years of human social development, the highest speed we could reach was about 20 kph – the pace of a sprinting hunter. This was more than doubled some 4,000 years ago with the mastery of horses. This same doubling occurred by the late nineteenth century with the development of trains and motor vehicles, multiplied a further three or four times within the following fifty years using aircraft and again with the advent of jets, and once more with the invention of spacecraft. At this rate, the speed exponential curve shows that it should be possible to reach 1% of the speed of light by 2070 and 5% by 2140.

The consequence of all this for our Ark voyagers is that they may arrive at their destination to find the planet colonized long before by their own race who travelled there in a fraction of the time.

A variation on this idea is long-term colonization. The physicist Frank Tipler has lent his support to the notion that an advanced race could 'planet-hop'. He bases the idea on the way the South Sea Islanders spread across the Pacific Ocean by island-hopping followed by consolidation. Using this model he believes there are two time factors to consider. Firstly, the time needed to make an interstellar journey (t_1), the other, the time needed to establish a colony and to prepare for the next hop (t_2). Conservative estimates for the journeys would be in the 1000–10 000 year range and a reasonable period for colonizing and consolidation would be in the region of one hundred generations.

Using this system, the galaxy can be completely colonized in a surprisingly short time, because the growth would be exponential.

If we say the journey time and the colonization time combined equal an average of 10 000 years, $t_1 + t_2 = 10\,000$; assuming there are 1 billion suitable planets in an average galaxy, these could all be reached and colonized in under 1 million years.

Figure 1.2

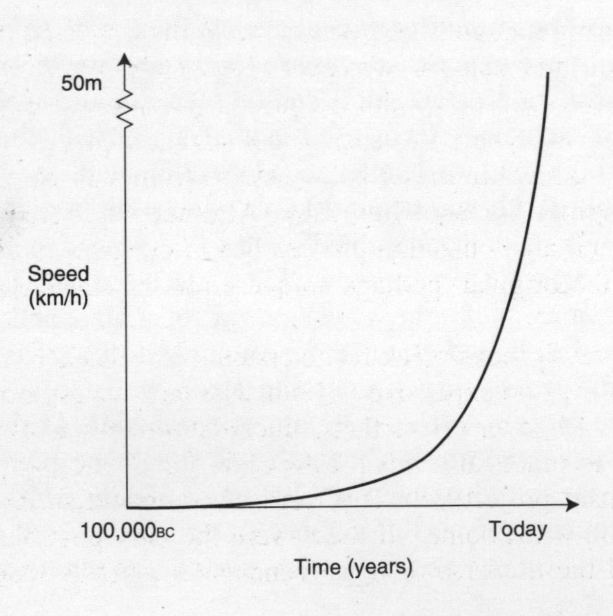

Figure 1.3

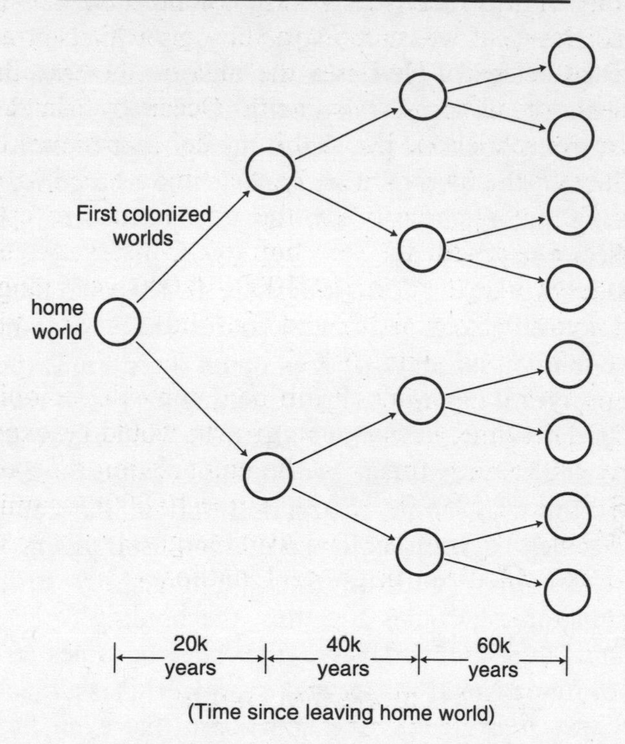

It is a sobering thought that perhaps the Earth was once the site of colonization and that for whatever reason the colony did not prosper and the 'wave' of colonization moved on, leaving us behind. If this was the case, such a colonization process would mean that all human life in our galaxy stems from a single mother planet, an original home of humanity. Alternatively, it could be argued that such a colonization process lies in our own future and that Earth is the original, perhaps unique home of *Homo sapiens*.

From this survey it looks bleak for interstellar travel. Each method is either too slow, too expensive or both. The best we can hope for is to develop antimatter drives that reduce the time for a relatively short journey to practical levels for the crew, but destroy any hope for an organized project with any form of command structure or communication with 'home'. If these were the only possible ways to get around the universe then the conclusion must be that alien races are not visiting us and no matter how advanced a civilization might become, interstellar travel will only be achievable on a very limited scale. Fortunately, there may be other ways around the problem using what I will call 'exotic physics'; that is concepts that do not break the laws of physics as we understand them, but merely bend the rules.

Our first piece of exotic physics is the concept of the wormhole. Like antimatter, the idea of wormholes arose as a consequence of manipulating the mathematics of physics, this time, Einstein's general theory of relativity.

Scientists have known for a long time that when a star has used almost all its available fuel it begins to die and the way in which it dies depends upon its mass. If it is about three times the mass of our Sun or larger it begins to shrink, setting up shock waves which result in an enormous explosion – the most violent event since the Big Bang – a supernova. But even then, because the Sun was so large to begin with, some material is left at the centre of the supernova which begins to collapse in upon itself again. This time, the matter becomes so dense that the incredibly strong forces holding subatomic particles together, the binding forces between quarks,* are overwhelmed, and the star becomes a seething cauldron of fundamental matter and energy. This is a black hole, so called because it is so massive and dense that even light cannot travel fast enough to escape its gravitational field.†

It often happens in science that mathematics tells us something should exist and what its properties should be before it is observed. Although the existence of black holes has not yet been confirmed, there are some promising candidates, and it is more than likely that they do exist somewhere in the universe. The chances of finding wormholes is slimmer, but there is nothing within the laws of physics that says they could not exist.

Einstein's general theory of relativity published in 1916 is an extension of the more limited special relativity. Special relativity was only concerned with observers moving at a constant velocity,

* The most fundamental form of matter known, constituents of protons, neutrons and electrons.

† By comparison, the shuttle has to achieve a speed of 11.18 kilometres per second to escape the gravitational pull of the Earth. Remember light travels at 300 000 kilometres per second.

but Einstein next wondered what the situation would be for objects experiencing acceleration. He imagined a lift in a state of free fall and a beam of light entering a hole in one wall. People in the lift would perceive the light travelling in a straight line. But to an observer outside the lift, the light would travel along a curved line. Einstein stated that this bending of light was caused by the fact that the lift was experiencing acceleration and he went on to say that because gravity is a form of acceleration, light would be bent by it.

Until Einstein, physicists saw the universe in three dimensions, with time as an extra factor. In general relativity, time is a dimension, just like length, breadth or depth; the universe actually exists in four dimensions called 'space-time'.

The only way we can visualize a four-dimensional universe is by representing it in three dimensions. Imagine a rubber sheet stretched flat. Now place a heavy ball in the middle – the sheet around the ball is misshapen the way space-time distorts around a massive object like a star. Roll a marble along the sheet near the heavy ball and it follows a curved path, just as light does near a star. A black hole is so massive and has such a powerful gravitational field, it curves space so much that within it lies what is called a 'singularity', a point at which the curvature of space-time becomes infinitely sharp and all the laws of physics break down. Wormholes, as theorized by a number of scientists (including the physicists who first postulated them – Kip Thorne and Michael Morris at Caltech in California), are created when two singularities 'find' each other and join up.

The reason wormholes are useful to interstellar travellers may be visualized from the diagram on the following page. Because of the nature of curved space-time, they offer a short cut, bypassing the need to travel between point A and point B using the conventional route.

Now, this is obviously an attractive idea and could eliminate all the problems faced by the near-light-speed traveller at a single stroke. But, as you would expect, there are problems with the method.

Firstly, wormholes are still pure speculation. They are not disallowed by the known laws of the universe, but neither are we certain they exist. But, assuming they do, they would probably be quite rare. The second problem is that it would be impossible to know which parts of the universe they linked until they were used. Furthermore, if they were usable, they would offer only a very limited service, linking the starting point to one fixed destination. It would be a bit like having a motorway connecting London with

Figure 1.4i

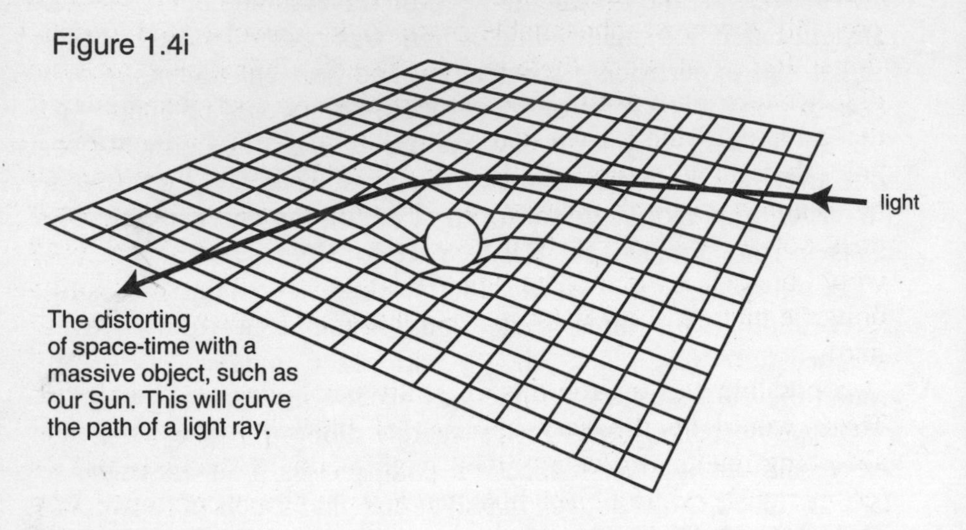

The distorting of space-time with a massive object, such as our Sun. This will curve the path of a light ray.

Figure 1.4ii

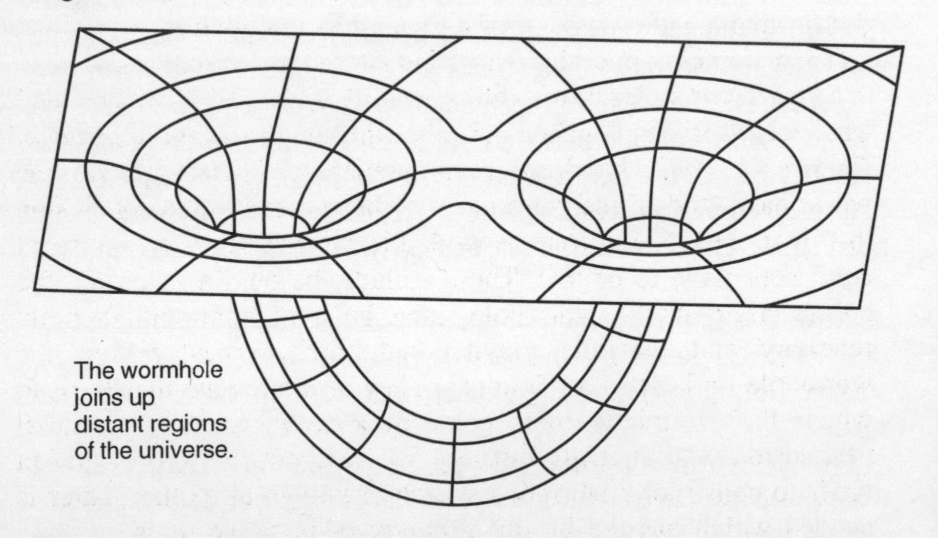

The wormhole joins up distant regions of the universe.

some other mystery location with no junctions or turn-offs *en route*.

Ignoring this drawback, we have to consider the nature of the link, and from what we know of black holes, a 'natural' wormhole would offer a very bumpy ride indeed. The inside of a black hole is probably the most inhospitable place in the universe; the gravitational forces at work there would instantly break any material object into a soup of fundamental particles and energy and even if these forces could be resisted, once within the grip of the black hole, there is no escape. So, the idea of using a wormhole created by joining two black holes at two different points in the universe does not seem very practical. The only way they could be used would be if there are certain types of black hole somewhere in the universe that do allow passage, but these might be very difficult to find.

A possible way around this difficulty is the idea of white holes. These would be the very opposite of black holes; rather than absorbing matter and energy they might act as perfect emitters or 'cosmic gushers'. If a black hole and a white hole were joined they could act as a one-way wormhole and circumvent the problem of escaping a black hole once it has been entered. Unfortunately, detailed mathematical analysis of this scenario has shown that such a system would be unstable and the white hole would rapidly decay, making the passage of a spaceship impossible.

There remains one alternative – man-made wormholes.

Since wormholes were first postulated by Kip Thorne and Michael Morris in a paper published in the *American Journal of Physics* in 1987, hundreds of theoretical physicists around the world have studied the concept. They have come to the conclusion that in order to construct a workable wormhole, a set of strict conditions have to be met. These include the obvious fact that the construction of the wormhole must be consistent with general relativity and that the gravitational 'tidal forces' within the wormhole be kept to a minimum. They also stipulate the shape to which the wormhole must conform and the mass of material needed to create it. Unfortunately, the mathematics shows that in order to construct a wormhole, material known as exotic matter is needed which has the bizarre property of negative mass.

Although wormhole enthusiasts insist that such a nonsensical idea can be realized within the laws of physics, most scientists dismiss the notion. If they are right it would appear that wormholes could never be manufactured, no matter how advanced a civilization might become. If they are wrong and the wormhole supporters are correct, then exotic matter has to be found and manipulated by the civilization before the wormhole could be built and used. (see chapter 10)

If wormholes look implausible, we are left with one other alternative method of interstellar travel which could facilitate a way around the light speed restriction, a concept made famous by the TV series *Star Trek* – the warp drive.

Science-fiction writers since the 1940s have bantered around expressions such as 'space warp' and 'hyperspace'; but, although many have been scientists themselves, they have rarely attempted to explain the concept in any detail. It has been visualized as the only possible way to circumvent the impracticalities of sub-light speed travel and the nuisance of having to work within the laws of physics, but what is warping and how could it be accomplished?

Another name for warping could be surfing. This is because it is based upon the principle of manipulating space-time itself so that the space vehicle moves on a 'wave'. The spacecraft would have the ability to alter space-time, so that it expanded behind the craft and contracted in front of it. This means that even though the craft is itself moving relatively slowly, the departure point would be 'pushed' back a vast distance and the destination 'drawn' nearer.

This sounds like a cheat but again it is a possibility within the rules of general relativity. The difficult aspect is once again the energy requirements. For the system to work, space-time would have to be distorted significantly (or else the effect would be so small that sub-light travel would probably be quicker).

Observation of our sun shows that its mass curves space-time so that it bends light by just one thousandth of a degree. For a spacecraft to utilize the expansion and contraction of space-time itself it would have to distort the space-time continuum much more than this. In some respects the vehicle would have to behave a little like a tiny black hole. Using this as a basis for calculating the energy requirements, the result sounds depressingly familiar. To make a black hole the size of a typical spaceship, say a disc 50 metres in diameter, we would need a mass of about 50 000 Earths

Figure 1.5

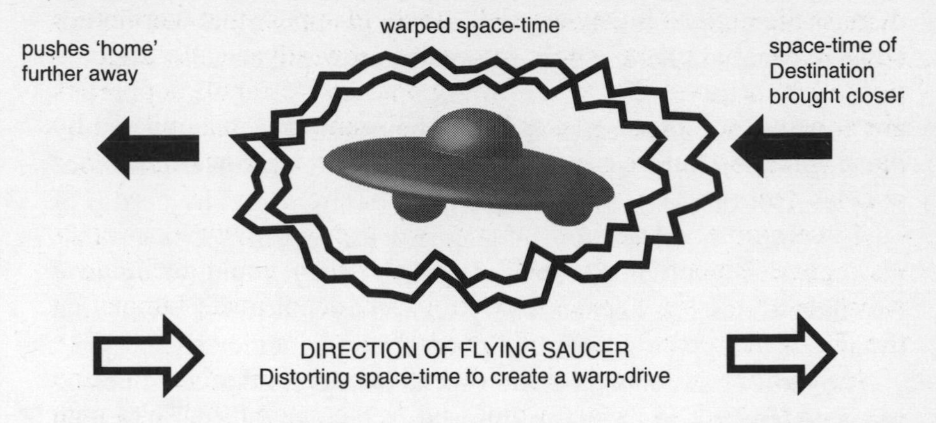

compacted into the space. Expressed in terms of energy this would be about equal to the entire output of the sun during its lifetime.

What then is to be concluded from these arguments? All forms of sub-light travel restrict meaningful interstellar travel and the options for circumventing the light-speed barrier present huge technical difficulties.

It may well be that alien intelligences have developed ways of producing enough energy to distort space-time or to create usable wormholes. To do this, their technology would need to be thousands of years in advance of our own, but, as we shall see in the next chapter, this is quite feasible.

Sceptics use the difficulties associated with interstellar travel as an argument against the possibility of alien visitors, but this is a narrow-minded approach and implies that other civilizations could not have developed faster or earlier than us. Far more damning are the pseudoscientific explanations for UFOs and alien visitations currently filling the internet and the news-stands, coming from enthusiasts themselves.

It is an interesting fact that descriptions of UFOs conform neatly to the historical period in which the observation is made. At the beginning of this century, witness descriptions of alien craft often bore a remarkable resemblance to airships and elaborately decorated flying machines not unlike something H.G. Wells would have described. Today, UFOs are supposed to use warp drives similar to those that power the fictitious *Starship Enterprise*. Yet hundreds of thousands of sightings are reported each year. Even if just one of these is genuine, there is a case for extraterrestrial travellers paying us a visit.

Clearly, if a culture can survive long enough they will eventually develop the technology to do almost anything within the laws of physics. But this isn't to say that we are being visited by hordes of aliens, some of whom are in league with organizations on this planet with intentions of subverting our society. Equally, the case for alien abduction appears badly flawed in spite of the fact that thousands of apparently normal people report incidents each year.

The most common explanation for abduction is that alien beings are a) interested in studying humans and b) involved in genetic experiments. The problem with these arguments is that any race advanced enough to manipulate space-time and travel across the galaxy would not need to conduct physical examinations or to physically extract genetic material. Putting aside the argument that any race so advanced would probably consider such behaviour immoral, both processes as described by abductees are ridiculously crude and sound suspiciously like the product of over-active but underdeveloped human imaginations.

I hope that one day our civilization will design and build interstellar space drives of one form or another and I am sure others already have. Alien races may have once visited a rather insignificant little planet called Earth and perhaps even pass this way from time to time, but notions of invasion forces or subversive three-foot tall grey beings with unfeasibly large eyes is suspect to say the least. It implies that the human race is in some way special and anyone who believes in life on other planets cannot also think in that way. In this sense, the beliefs of some UFO enthusiasts differ little from the egocentric fallacies of any other organized religion.

Chapter 2: Is There Anybody Out There?

Alone, alone, all, all alone.
Alone on a wide wide sea!

Samuel Taylor Coleridge

Even though we cannot yet build interstellar spaceships, we may look at the stars and wonder: are we alone in an infinite universe or is the cosmos as full of life as the Earth is? This has been perhaps the biggest question facing our species since we developed the ability to think beyond our material requirements and even now, we are only edging slowly towards an answer.

Scientists know there is certainly life on one planet – the Earth. But because we have only this one example upon which to build hypotheses, knowing for sure whether the series of events leading to life here is unique or extremely common is impossible – until we have more evidence. And, because of the almost unimaginable distances between stars, it is only now that we can travel outside our own planetary atmosphere and develop machines that can see into the deepest recesses of space that we can begin to hope for conclusions.

It has been known for most of this century that life on other planets within our neighbourhood, our Solar System, is extremely unlikely. Of the two planets nearest to the Sun, Mercury and Venus, Mercury has extremes of temperature with part of its surface an inferno and the other a frozen wasteland, whilst Venus has surface temperatures of some 800K. Beyond the third planet, Earth, lies Mars, once thought to be the most likely contender for extraterrestrial life within our Solar System. In fact as recently as 1877, the astronomer Giovanni Schiaparelli created a flurry of excitement by announcing that he had observed a network of what he called *canali* on the surface of the planet. *Canali* was wrongly translated into English as 'canal' instead of its true meaning of 'channel' and astronomers all over the world began to see increasingly complex canal systems as the rumours spread. Sadly, although the news inspired H.G. Wells to write *The War of The Worlds*, there are no canals on Mars; the effect was produced by a natural coloration of the surface.

The Viking probes of the 1970s found no trace of life on Mars, not a single microbe, but more recently, in August 1996, NASA scientists announced that a sample of rock from a meteorite – sample ALH 84001 – might contain fossilised martian microbes. At the time of writing, this sample is undergoing further tests by NASA scientists.

Beyond Mars lay the gas giants, Jupiter and Saturn, with atmospheres containing toxic gases constantly churned up by powerful magnetic fields. The two largest satellites of the Solar System, Titan, orbiting Saturn and Ganymede, Jupiter's largest moon, might be more promising, and the Voyager probes that have passed close by have found what is believed to be organic molecules on the surface of Titan. But with surface temperatures around –150 °C (123K) and toxic atmospheres with little trace of oxygen, the chances of these molecules developing into living matter is very small.

At the outer edge of the Solar System, Uranus, Neptune and Pluto offer little comfort for what scientists refer to as carbon-based life-forms because again the temperatures are either too low, their atmospheres noxious or, in the case of Uranus, the entire planet is a single ocean of superheated water, warmed by volcanic action and covered in poisonous gases.

To find life, particularly life we can readily recognize, we must turn our thoughts and our telescopes and probes beyond the tiny confines of our Solar System to the distant stars. But the problem we face then is distance. Our Solar System is vast by everyday scales, some 12 billion km across, but it becomes a meaningless speck and such numbers trivial when we begin to imagine contacting beings that live on planets orbiting other stars.

The nearest star other than our own sun is Proxima Centauri, which lies 4.2 light years from Earth. As we saw in Chapter 1, this is a staggering distance. What this means is that light, which travels at just over 300 000 km/sec would take 4.2 years to get here. This is equivalent to a distance of 300 000 × the number of seconds in 1 hour (3600) × the number of hours in 1 day (24) × the number of days in 1 year (365) × 4.2, which comes to a little under 4×10^{13} kms (4 with 13 noughts after it, or 40 million million kilometres). This is roughly equal to 100 million trips on an Apollo spacecraft to the Moon. At the speed Apollo capsules travelled (about 40 000 km/p/h) it would take about 100 000 years to cover the distance to even this, our nearest neighbour.

So, until we develop the technology to improve our speed, we can only hope to a) contact aliens using light-speed signals such as radio waves, b) wait for them to contact us, or c) use telescopes of various types to try to discover as much as possible about other planets we may find orbiting nearby stars.

But just what are the chances of there being intelligent life beyond our own world?

Scientific opinion is split. There are those, like the astronomer Frank Drake, creator of the first SETI (the Search for Extraterrestrial Intelligence) project, or scientist and author Carl Sagan at Cornell University, who believe the universe is teeming with life. At the other end of the spectrum, there are writers and pundits, such as Marshall Savage, author of the *Millennial Project* and the physicist Frank Tipler who think we are totally alone.

The problem with trying to come up with any form of definitive answer or even an approximation is that we have no clear idea of all the variables to be considered or how these interrelate. For example: How likely is it that molecules of DNA can form given a long enough time period? How frequently do planets form around stars? How likely is it that even complex molecules can evolve into living material? We know all these things have happened at least once, but has it been only once, or billions of times?

To try to quantify the argument, the pioneer in the search for extraterrestrial intelligence, Frank Drake produced in 1961 a now famous formula which has since become known as the Drake Equation. It is very straightforward and a surprisingly powerful tool for the astronomer, except that almost all the variables can show a range of values, and no one is yet sure what numbers to put in. It is the work of astronomers, biologists and geologists to gradually narrow down each of those numbers to something more workable and to then come up with some form of answer to the Drake Equation.

The equation is:

$$\mathbf{N} = \mathbf{R} \times \mathbf{f_p} \times \mathbf{n_e} \times \mathbf{f_l} \times \mathbf{f_i} \times \mathbf{f_c} \times \mathbf{L}$$

Although this might look daunting, it is actually as easy to use as working out your expenses. The letter **N** signifies the number of civilizations in our galaxy trying to make contact. Each of the symbols on the right hand side of the equation represent separate factors which have to be considered in addressing the question: Is there life beyond Earth? (Each term is considered in isolation, in other words the number assigned to say $\mathbf{f_p}$ is independent of that given to **L**, $\mathbf{f_i}$ or any of the others). When numbers for all of these factors are plugged in, we end up with a figure for **N**. So what are these factors?

First, **R**; this stands for the average rate of star formation. A common misconception is that the universe was made at the time of the Big Bang and that was it, no change ever since. Of course this is not the case. The prevailing theory is that the universe is expanding and stars and planets are being created and dying constantly. Scientists are beginning to actually see this birth process using instruments such as the Hubble Space Telescope. It seems that some parts of the galaxy are more fertile than others and the process of star birth is far slower than it was at distant points in our galaxy's past, but at a conservative estimate, astronomers think that about 10 new stars are formed in our galaxy every year. So **R** is one of the variables which is pretty much agreed upon, that is 10.

$\mathbf{f_p}$ is the fraction of stars that are 'good' and could contain planetary systems. By 'good' astronomers mean suitable for forming and keeping stable, Earth-like planets in orbit around them. This is a rather complex matter. The age of the star must fall into a certain range. If it is too old, its fuel will be running down and it will emit radiation that would be unhelpful for the formation and sustainment of carbon-based life. Also, as a star gets older, the rotation of planets in orbit around it begin to slow. If the star is more than around 6 billion years old, (our sun is about 5 billion years old), this will have a dramatic effect. Planets orbiting very old stars will have stopped rotating altogether and will have one face permanently turned towards the sun, and the other existing in permanent night. If the star is too young, it may not have had time to allow planet formation and the mechanism that creates and evolves life-forms to run its course.

More importantly, planets that can sustain life-forms capable of developing civilizations could not be found orbiting pulsars or quasars – exotic stellar objects which would emit forms of damaging radiation – nor could the home star be unstable over long time periods.

Finally, many stars are binary – that is they are made up of two stars orbiting one another. Although this system by no means rules out the formation of planets, binary stars are generally considered less likely to possess Sol-like systems than single stars.*

When Drake first suggested his equation, the value for $\mathbf{f_p}$ could only be guessed at, but recently astronomical findings have begun to narrow down the range of numbers this could be. Back in the early 1960s, Drake placed $\mathbf{f_p}$ at about 0.5, in other words, half the number of stars in the galaxy were potentially able to form planets, but then, when observational techniques improved and new data was gathered, the initial results showed this figure to be wildly optimistic.

Using present-day technology, planets orbiting other stars cannot be seen in the way astronomers observe the planets in our own Solar System; the distances, as we have seen, are simply too great. It has been estimated that a telescope the size of the moon would be needed to observe clouds and coastlines on a planet within 50 light years of Earth. And that is not the only problem. Imagine trying to detect the presence of a firefly perched on the edge of a spotlight from a few hundred miles away. The light from the spotlight would completely swamp the effect of the firefly. In the same way, the light from an orbiting planet, which is merely reflecting light from its sun, would be totally overwhelmed by the far greater magnitude of the star.

These two problems would, you might think, stop us ever knowing if any star other than our own possesses planets, but there are other ways of knowing if a planet orbits a distant star.

The best technique we have today is observing 'wobble'. If you can picture a hammer thrower at the Olympic Games spinning on the block and just set to let go of the hammer, the athlete, who is substantially heavier than the hammer, has a greater pull on the chain and the hammer, but the hammer (which weighs about 7 kg) also has a pull on the thrower who might weigh around 20 times as

* Sol is the name given to our sun.

much. With suitable instruments this pull, or 'wobble' could be measured. In the same way, a planet in orbit around a star will exert a pull on the star in an identical but much smaller way than the star pulls the planet towards it. Obviously, the bigger the planet, the greater the effect.

Even so, this is a very sensitive technique and the difficulty of measuring the wobble of a star has been compared to using a telescope on Earth to see a man waving on the moon. Yet within the past two years a refinement of this procedure has been used successfully by astronomers based in Geneva in Switzerland – making them the first to confirm the existence of an alien planet.

The discovery of a planet orbiting the star 51 Peg in the constellation of Pegasus, announced in October 1995 at a conference in Florence, startled the astronomy community and made headlines around the world. The planet (which is still unnamed) is half the mass of Jupiter (the largest planet in our Solar System and about 300 times larger than the Earth) and orbits the star at a distance of about 8 million kilometres.

Although most astronomers believe it unlikely that gas giants like Jupiter could sustain carbon-based life, it is suspected that solar systems that have any chance of containing an Earth-like planet need at least one Jupiter-type planet which would probably be found further away from the star than the region in which solid, cooler planets would be located. The reason for this is that gas giants act like vacuum cleaners, soaking up stray asteroids, comets and meteors that enter the system. In this way they protect the inner, Earth-like planets facilitating the chance of a stable environment within which life could form and a civilization develop. However, the planet found around 51 Peg is in the wrong place. It is too close to the star (Mercury orbits between 46–70 million kms from the Sun) and it circles the star too rapidly, taking only 4 days to complete an orbit (Mercury takes 88 days). But it is a planet and the discovery has created a revolution in our way of thinking about the cosmos. We now know our Solar System is definitely not the only one.

Within months of the discovery of the planet orbiting 51 Peg, more solar systems were discovered. In January 1996, two new planets were found around different stars, one orbiting 70 Virginis in the constellation of Virgo and the other in the constellation of

Figure 2.1

Our Solar System

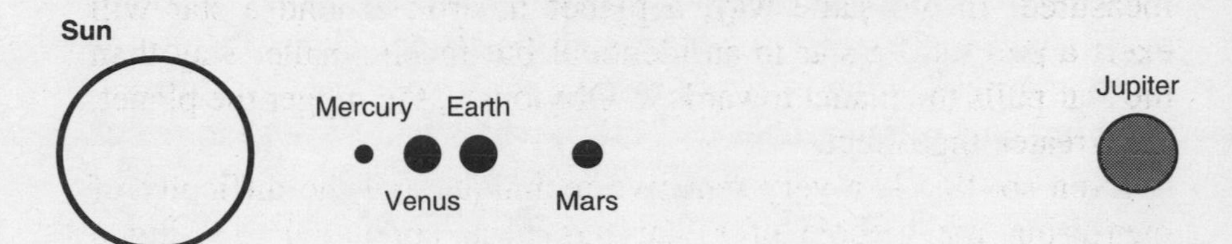

New solar systems

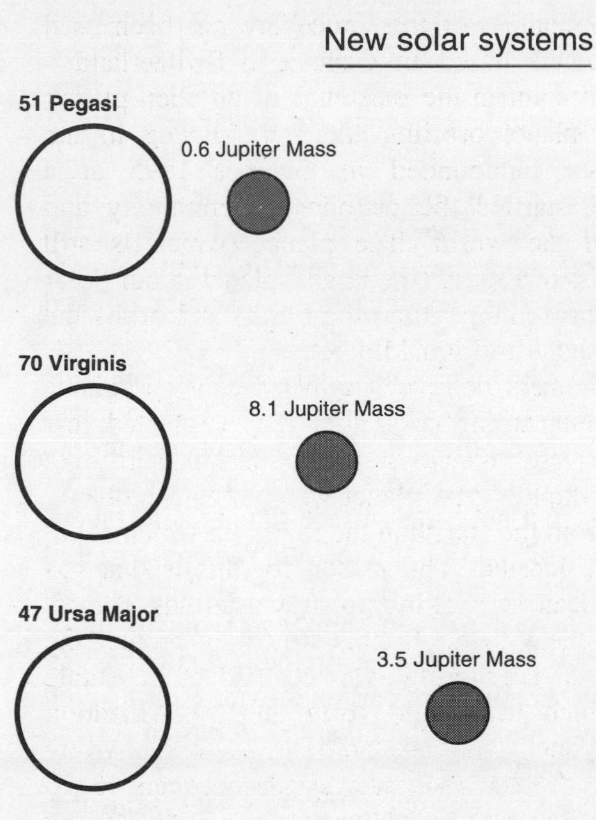

Comparison of Planets

	Earth	Jupiter	51 Pegasi	70 Virginis B	Ursa Majoris B
Mass	0.003	1.00	0.6	8.1	3.5
Diameter	0.09	1.00	0.3–1.30	0.3–1.00	0.3–1.1
Distance from star	1.00	5.20	0.05	0.43	2.1
Day temp (°C)	15	-150	1000	85	-80

Ursa Major, a star named 47 UMa. Both of these are around 35 light years from Earth. Both are Jupiter-type planets, but these are in the 'conventional' position around their stars. Since then, four more planets orbiting different distant stars have been discovered, including one orbiting Lalande 21185, which is the 4th nearest star to our sun.

So what do these findings mean for the number we assign to $\mathbf{f_p}$? The search for stars containing planets using the technique of 'wobble' detection has so far spanned over a decade and involved choosing hundreds of Sol-type stars. Until these recent discoveries, none of the stars gave positive results and astronomers were beginning to despair of ever finding a candidate. So, although these recent finds have greatly encouraged astronomers, they have also shown how the initial guess was too high and at the same time illustrated how difficult it is to assign numbers to the Drake Equation. Instead of 0.5, a more conservative estimate might now be 0.1, or that one in ten stars are capable of developing and sustaining a planetary system.*

Next we come to $\mathbf{n_e}$, which is the number of planets per star that are Earth-like. Once again, we are bound by very limited experience. In our solar system, there is really only one Earth-like planet. Mars may once have possessed a more conducive atmosphere than it does today, but now the surface temperature ranges from – 50 °C (223K) to 0 °C (273K) and the atmosphere is so thin (with an atmospheric pressure about $\frac{1}{100}$ of Earth's) that humans would need to carry their own oxygen supply to work or move around on the surface. At the other extreme, Venus has an atmosphere made up almost entirely of carbon dioxide (CO_2) which produces such a severe greenhouse effect that the Venusian surface is hotter than that of Mercury.

So, again being conservative, let us put $\mathbf{n_e}$ for our solar system at 1.

$\mathbf{f_l}$ in the Drake formula stands for the fraction of Earth-like planets upon which life could develop and in trying to assign a number for this we really are in an almost totally unpredictable domain.

Firstly, what is life? It might seem an obvious question but

* A caveat to this is of course that the technique can only detect large planets and the theory that planetary systems invariably contain gas giants may be wrong, so the figure may be higher.

biologists are quick to point out that the standard parameters are open to debate. Living beings grow and move, but so do crystals, producing regular patterns made of repeated simple units much like cells, and inanimate water or any other liquid can flow, or move. Life-forms use energy, but so do computers, trains, rockets. Perhaps a better definition would be that a living being can *control* energy.

An alternative could be to argue that only living things process and store information, but this is the sole purpose of a computer. The debate about the possible future development of intelligence by complex computers still rages, yet the desktop computer I'm using to write this could not be described as living. So, what other criteria could we use? Could the ability to reproduce constitute life? But flames reproduce. Probably the best definition would be that life-forms reproduce and pass on genetic material, inherited characteristics, to their offspring.

For the purposes of this book, I'm really interested in reaching a conclusion about intelligent life-forms with which we can communicate readily. It may be that any number of exotic creatures live in this almost infinite universe, but the chances of contacting them or communicating with them is even less likely than the probability of encountering a life-form with which we can communicate. There is even the possibility that we have encountered these beings and have, for one reason or another, been totally unaware of them or they of us. So, to allow for 'life as we know it' we need to think in terms of carbon-based life able to communicate with us. But why carbon-based?

According to the laws of physics (which in turn give us the rules of chemistry and subsequently biology), carbon is the only element which can form complex molecules, known as organic molecules. For 'life as we know it', the alien beings must operate naturally within the same narrow limitations of temperature, pressure and radiation as we do, not least because if they did not, we would probably find it impossible to communicate with them. Within those parameters the only element to form organic molecules that constitute cells, tissue and flesh is carbon. Carbon has the unique ability to form very strong inter-atomic bonds with a large number of elements such as nitrogen, oxygen and hydrogen as well as multiple bonds with other atoms of carbon. This allows it to form a vast range of molecules, some of which can contain tens of thousands of atoms. No other element can approach this level of versatility.

For a planet to be the home of carbon-based life it had to possess a certain set of environmental conditions and materials in its primeval history and a subsequent set of finely tuned conditions and materials for life to have evolved and flourished there.

Sceptics argue that these conditions are unlikely to be duplicated in the universe and that the probability of life evolving is therefore slight, but there is a growing body of evidence to oppose this view.

In 1953, just as the structure of DNA was being elucidated by Crick and Watson in Cambridge, two scientists at the University of Chicago, Stanley Miller and Harold Urey, were investigating the initial conditions on Earth that produced the biochemical environment in which life began. It was known that life had originated on Earth a little under 4 billion years ago and that the predominant chemicals in the atmosphere at that time were ammonia, water and methane. Miller and Urey placed these chemicals in a jar and allowed an electrical discharge to pass through the mixture. After sustaining this process for several days they found a red-brown deposit had formed at the bottom of the jar. When this 'primeval soup' was analysed they found it to contain amino acids – organic molecules which act as the building blocks for all life on Earth.

Further experiments showed that a wide range of molecules essential for life could be formed in this way. The pair added another simple molecule, hydrogen cyanide (HCN), found in volcanic gases, and discovered that a number of complex molecules which are key to the formation of proteins and DNA were formed. Although these molecules are a long way from the giant structures of DNA (deoxyribonucleic acid) and RNA (ribonucleic acid), the elaborate molecules that encode the production of proteins and the day-to-day workings of cells, Miller and Urey postulated that a 'soup' of life-forming molecules could have been brewed in the Earth's atmosphere during the space of just a few years. Recently, Stanley Miller has declared that these molecules could have developed in complexity and produced living cells within perhaps as little as 10 000 years. Flying in the face of critics who claim life on Earth is unique, based on his own experiments,

Miller is sure that given the correct environmental conditions and the proper blend of chemicals, life could form on any planet.

Finally, the fossil record shows us that life began on Earth at the earliest opportunity. In 1980, fossils of creatures called stromatolites or 'living rocks' were found in the Australian desert which were the simplest and probably the most ancient form of life on Earth, living over 3.5 billion years ago. We know the environmental conditions for life only became suitable around 4 billion years ago and so it would seem that a period perhaps of only a few hundred million years passed before the very simplest life forms began to appear. This does not provide evidence for the evolution of life on planets other than the Earth, but it illustrates the notion that given the correct conditions, life will appear readily. The great proponent of extraterrestrial life, Carl Sagan has written: 'The available evidence strongly suggests that the origin of life should occur given the initial conditions and a billion years of evolutionary time. The origin of life on suitable planets seems built into the chemistry of the universe.'[1] What he means by this is the principle of self-organization.

In recent years it has been suggested that certain physical and chemical systems can leap spontaneously from relatively simple states to ones of greater complexity or organization. This organizing principle, some argue, is a form of anti-entropy effect which may be, in some mysterious way, linked to life itself. Entropy is the 'level of disorder' in a system and in nature entropy always increases – an apple left to stand will gradually decompose, its cells break down and the 'neat', 'organized' form of the fresh apple will decay into a disorganized mush. The self-organization principle could, it is believed, help to reverse the natural tendency for entropy to increase in the universe. As a consequence the chances of life deriving from a collection of complex organic molecules is greatly increased.

According to Frank Drake, 'Where life could appear, it would appear.'[2] He assigns a value of 1 to the parameter $\mathbf{f}_l$. In other words, there is a 100% chance that a suitable planet will form life if it has the correct conditions. Others, such as the Nobel Prize-winning chemist Melvin Calvin and Carl Sagan have concurred, believing that life is more likely than not to form on a suitable planet.[3] For others, those who do not believe in extraterrestrial life, the value for $\mathbf{f}_l$ is the most crucial of all the terms in the Drake Equation. They place it at 0 which would consequently make **N** equal to 0 – no life anywhere else in the universe. The value for $\mathbf{f}_l$ probably cannot be anything other than 1 or 0, so, for the purposes of our discussion, I will give it a value of 1.

Next we turn our attention to $\mathbf{f}_i$. This is the term that refers to the fraction of Earth-like planets where life has become intelligent. Again, when we first contemplate this expression, we are struck by the need to define. This time the question is: What constitutes intelligent life?

Many would argue that dolphins and whales are highly intelligent animals that could have formed a civilization if they had evolved on land. They can communicate with members of their own species and have been known to interact in a highly intelligent fashion with humans. Attempts have even been made to decipher the complex sequence of clicks and squeaks they use to communicate with one another.*

In a different sense, ants and bees act in an intelligent fashion when considered as a collection of individuals, each acting as a unit in a larger society, a Gestalt. So, if we were to apply Drake's Equation to our planet, we could arrive at a value for $\mathbf{f}_i$ between 1 and at least 4, but again being conservative, let us take the value of 1, representing the human race.

The penultimate term, $\mathbf{f}_c$, represents the number of intelligent species who would want to communicate with us and again, we are faced with a highly subjective parameter for our equation. In order to use this term we have to place some limitations upon how we arrive at a value. We must first assume that an intelligent race uses some form of electromagnetic radiation with which to communicate and to interact with their universe. Most scientists would agree that it would be unlikely for an intelligent species to have developed without using any form of electromagnetic radiation. An alien intelligence may utilize extreme regions of the spectrum, they may see in the infrared or the ultraviolet because of the nature of the light emitted by their sun. Alternatively, they may live in

* Interestingly, these efforts have so far met with only limited success which may act as a salutary lesson for those attempting to contact alien civilizations on distant planets. Supposing we eventually do make contact, how likely is it that the two races will ever understand one another?

extreme conditions such that vision is as unimportant to them as it is to some deep-sea creatures, but whatever extreme situation that may be imagined, they must use some form of electromagnetic radiation. If this is not the case, then such an alien race would fall out of the category of 'life as we know it'.

As a civilization, we utilize a range of radiation, from radio and television signals to X-rays, from ultrasound to microwaves, so it is likely that any civilization at least as advanced as us would also employ similar electromagnetic waves within their technology; they may even have developed something similar to television or radio. Even if they had not created an entertainment system which leaked signals into space as our televisions have done in recent decades but were actively interested in communicating, they should be able to build equipment that would receive and decipher signals from space.

This then leads to a question concerning the sociological and psychological make-up of an alien intelligence: Would they necessarily want to communicate with us? It is a serious point that the signals we have been sending inadvertently into space may have presented our race in a very poor light. For some fifty years our calling card has been television signals conveying images of everything from the most violent Hollywood films to news coverage of war, famine and torture, leaking into space in far greater quantities than any form of contrived, politically correct message we may wish to send to our celestial neighbours. Many of these signals would be too weak to reach distant stars, but this is perhaps underestimating the sensitivity of alien detection systems. Television signals are no different to any other forms of electromagnetic radiation in that they travel at the speed of light. It is therefore conceivable that alien civilizations living on planets up to 50 light years away could be chuckling at our antics, or perhaps battening down the hatches for fear we'll ruin the neighbourhood.

So, what value do we give $\mathbf{f}_c$? On the one hand, it would seem likely that any civilization would eventually develop a form of long-distance receiving and transmitting system using electromagnetic radiation, enabling communication. But how many races would want to make contact? There could be an abundance of races busily communicating with one another but excluding us; equally, alien civilizations could prefer to keep themselves to themselves whether or not they have been warned off. Weighing up these factors, a conservative estimate would be that 10–20% of intelligent aliens would want to communicate, so $\mathbf{f}_c$ would be say 0.1.

Finally, we come to **L**, which represents the lifetime of a civilization (in years). And again, in attempting to assign a value, we face another complex series of permutations. The value of **L** puts us out on a limb, for we have to consider hypothetical sociological factors for a hypothetical race, but we have again one example to draw upon – our own experience.

It suggests an interesting synchronicity that our race developed weapons of mass destruction at almost the exact point we revealed ourselves to the universe with our electromagnetic signals, and it could be that many races are destroyed at the very point they could make contact with their neighbours.

Since Frank Drake first devised his formula in 1961, there has been much debate amongst scientists about the value of **L** and during the past 35 years the political and social *Zeitgeist* has altered radically. The Cold War has ended, but the threat of nuclear destruction is still very much with us and the killer instinct of humankind has not changed in the slightest. Perhaps the ease with which human beings make war is irreducibly linked with our drive to progress and advance. It is possible the instinct that drives us to communicate derives from the same source as our aggression. If this is the case, it may be no different with other species; it may even be a natural law. In such an event, it would be likely that a large proportion of civilizations destroy themselves at around the time they develop the technology to communicate beyond their own world.

There are also a number of other ways in which a civilization based on a single planet can be destroyed. Scientists are only now beginning to realize the very real danger of planetary collisions with comets or asteroids. It is believed that a devastating asteroid collision caused a sudden traumatic alteration in the ecosystem of the Earth some 65 million years ago resulting in the extinction of the dinosaurs and there have even been a number of documented near-Earth collisions this century. The massive explosion reported in Siberia in 1908 which devastated hundreds of square kilometres of forest in the region of Tunguska is thought to have been caused by a meteorite exploding several kilometres above the ground.* If this had occurred over London, millions would have died. An object only a few times larger than the Tunguska fireball impacting with the Earth would not only devastate a wider area, but the dust thrown up by the collision could produce a blanket around the entire planet capable of destroying all life on the surface. If it landed in the sea, the tidal effects would be almost as damaging.

There is also the question of planetary resources. We as a race are perilously close to over-exploiting the resources of our world and we are already capable of severely damaging planetary mechanisms that are there to maintain an ecological balance. It is conceivable that other civilizations have followed the same path and gone further, completely destroying their own environments. Such threats as reduced fertility, Aids, superbugs and nuclear terrorism are all further potential civilization-killers.

One conclusion to be drawn from these considerations is that civilizations either survive little more than 1–2 000 years or else they continue for perhaps hundreds of millennia. It is possible that many races pass through a 'danger zone' during which they have a high chance of destroying themselves, but if they come through it, they develop into highly advanced cultures capable of interstellar travel and colonization.

In an extreme case **L** also depends upon astronomical factors. If we assume that life may form on a large number of planets and that those life-forms could evolve into intelligent civilized beings, the time at which life began on their world would be a crucial consideration.

The universe is believed to be approximately 12 billion years old, and our sun is a very typical star located some two-thirds of the way along one of the spiral arms of the Milky Way galaxy – itself an 'ordinary' galaxy amongst an estimated 100 billion others. In astronomical and geological terms, the Earth is quite average and life began to appear here 4 billion years ago or around 8 billion years after the Big Bang. But it is quite conceivable that a great many planets around other older stars cooled long before our own planet. Astronomers have observed the death of stars far more ancient than our own. If any planet around these stars had brought

* Although some UFO enthusiasts offer the alternative theory that it resulted from the explosion of a spacecraft in the Earth's atmosphere.

forth life, any civilization that formed there would either be ancient, interstellar voyagers or long dead.

To find a sensible value for **L**, we must assume a normal distribution of ages for successful civilizations. If **L** for a particular planet is 2 000, the race may have destroyed itself and is of no further interest, but **L** could be much larger. It is possible there have been and still are civilizations hundreds of millions of years old. Equally there would be a large number of very young civilizations, perhaps no more than 2–3 000 years old. Most civilizations that have survived and are able to communicate would be somewhere between the two extremes.

Drake and his colleagues have placed a value of 100 000 on **L**. This seems rather arbitrary but nevertheless, if we put in any figure over the 2 000 year watershed the equation gives us a correspondingly large number for **N**, the number of advanced civilizations wanting to make contact.

We assigned **R** as 10, $\mathbf{f_p} = 0.1$, $\mathbf{n_e} = 1$, $\mathbf{f_l} = 1$, $\mathbf{f_i} = 1$, $\mathbf{f_c} = 0.1$, **L** = a large number. Now if we put these figures into the Drake Equation, we reach a very simple conclusion:

N = 10 × 0.1 × 1 × 1 × 1 × 0.1 × A large number.

The 10 and the 0.1 cancel out giving us just: 0.1 × A large number.

If we call **L** (the average age of a civilization) say, 100 000, it means there are 10 000 civilizations sharing just this single galaxy (one of 100 billion remember). Frank Drake believes the value of **L** to be much greater, which would mean that **N** would be correspondingly larger, and, according to some enthusiasts, **N** could be in the tens or even hundreds of millions. This may seem excessive, but when we consider that our galaxy contains upwards of 100 billion stars, an estimate of 100 million civilizations means that there is only one such race for every 1 000 stars.

So, how likely is it that we will communicate with one of these civilizations? Thus far every attempt to make contact has failed utterly; but that has not been for the want of trying. The first serious scientific effort to listen in on a possible alien dialogue (or monologue) came from a paper published in the science journal *Nature* in 1959. It was written by an Italian astronomer Giuseppe Cocconi and American physicist Philip Morrison who reasoned that if aliens wanted other species to contact them, they would make it as easy as possible for them to do so. What they meant by this was that an alien intelligence would broadcast a signal that would have universal meaning and be within the range that would be transmittable and detectable via radio telescopes.

Cocconi and Morrison chose as their standard the frequency 1.420 GHz (gigahertz). The reason for this choice is that 1.420 GHz is the frequency at which the element hydrogen is known to resonate. Because hydrogen is by far the most common element in the universe, it was assumed an alien intelligence which had developed radio technology would know of this fact and expect a recipient to be equally knowledgeable.

Enthusiasts immediately began scanning the sky calibrating their radio telescopes to this frequency. The first SETI project was led in the early days by Frank Drake who in 1960 established a team based at Green Bank, West Virginia in the United States. This was followed soon after by another SETI project created at a giant radio telescope centre called Big Ear, in Ohio, directed by astronomer Bob Dixon, who has been searching the skies ever since.

Today there are a number of SETI projects running concurrently around the world. NASA created a $10 million a year programme in 1992, approaching the task from a different angle to the early experimenters. Whereas Drake and others had focused on the frequency of 1.420 GHz and a select group of prime candidate stars within 50 light years of Earth, NASA decided to blitz the heavens with a broad sweep of all likely stars and a wide range of frequencies. Sadly, we will never know if the project would have garnered conclusive results because, little more than a year after it was established, funding was terminated by the interference of congressmen who thought they were paying for the indulgence of UFO cranks. One prominent member of Congress was reported to have said:

> Of course there are flying saucers and advanced civilizations in outer space. But we don't need to spend millions to find these rascally creatures. We need only 75 cents to buy a tabloid at the local supermarket. Conclusive evidence of these crafty critters can be found at check out counters from coast to coast.[4]

Such ignorance does nothing to help the serious search for life beyond our planet, but luckily there are others who have a more open minded approach. Rather aptly, Steven Spielberg, producer of *ET* and *Close Encounters of the Third Kind* is funding a project based on the East Coast of the United States and other wealthy enthusiasts are putting money into the search at a variety of sites around the world. The original NASA SETI project was scrapped but picked up by others with private funding and only last year Project Phoenix was initiated. This is a pan-global operation using the world's largest radio telescopes to sweep the universe between 1 and 10 GHz which happens to be the region where there are many natural resonances and emissions, (including that for hydrogen). NASA have also developed what they call spectrum analysers to search across wide ranges of frequencies within the chosen limits and are currently devising software to filter out noise and other interference in the signal.

Yet, despite the money and advanced technology, after almost forty years of searching, no conclusive evidence for what have been dubbed LGMs (Little Green Men) has been found. One famous false alarm in 1967 illustrates the difficulties facing astronomers in their search.

A PhD student at Cambridge University named Jocelyn Bell detected a strong, regular signal coming from deep space in the region of the spectrum then thought most likely to contain a contact frequency. After reporting the finding to her supervisor, Anthony Hewish, they agreed they would not go public until they had investigated the signal fully. Gradually they eliminated all possible sources both terrestrial and celestial until they realized that the signal was actually an emission from a strange object in deep space which was sending out an almost perfectly regular pulse. The object was then found to be a neutron star, or pulsar, the remains of a dead star that had collapsed under its own gravitational field so much that the electrons orbiting the nucleus of the atoms making up the star had been jammed into the nuclei and had fused with protons to form neutrons. This super-dense matter emits pulses with such regularity that pulsars are thought to be the most accurate clocks in the universe.

Since the discovery of Bell and Hewish, other regular signals have been detected which have not originated from pulsars or any terrestrial source, but have appeared only once. A team lead by

Professor Michael Horowitz at Harvard University has reported 37 such signals during the past 10 years all within 25 light years of Earth, but because they have not been repeated they do not qualify as genuine signals from a race trying to contact us. They could of course be one-off leakages from specific events, but we might never know and for scientists to analyse a signal properly, they need a repeated, strong, regular signal.

The search goes on and with the recent confirmation that other stars possess planets, some of the thunder has been taken from cynical doubters such as the Congressmen who saw no value in the funding of SETI. If there is life on other planets, which would seem more likely than not, we do have a chance of contacting them one day. If 1 in every 1 000 stars have planets which are homes to advanced civilizations, we should have at least one such star within 50 light years of Earth. Perhaps we are sending signals at the wrong frequency and tuning our instruments to the wrong region for that particular civilization. Maybe, we are unlucky enough to be neighbours to a race who either do not want to be contacted or have developed a technology that is so different to ours that we cannot yet contact one another. It could also be that signals sent from more distant planets have not reached here yet.

Sceptics quote what has become known as Fermi's Paradox after the famous Italian physicist Enrico Fermi who declared in 1943 that if aliens existed they would be here. But this is not only a tautological argument, it demonstrates staggering arrogance, assuming that it is a simple matter to detect alien life or that we are so important that aliens could want nothing other than to contact us.

Whatever the future holds for the efforts of scientists to contact alien beings, the effect of such a discovery upon our society and the mind set of each of us would be enormous. As the physicist Paul Davies has said:

> There is little doubt that even the discovery of a single extraterrestrial microbe, if it should be shown to have evolved independently of life on Earth, would drastically alter our world view and change our society as profoundly as the Copernican and Darwinian revolution.[5]

Although this is undoubtedly true, proof that intelligent beings have visited or are still visiting us here would be exponentially more significant. In fact, it would be the biggest news story in history, and might, just might be waiting for us around the next corner.

References

1. As quoted in, Paul Davies, *Are We Alone?*, Penguin Books (Orion Productions), 1995, p. 23.
2. Frank Drake and Dava Sobel, *Is Anyone Out There?*, Souvenir Press, 1993, p. 56.
3. Ibid.
4. Adrian Berry, *The Next 500 Years: Life In The Coming Millennium*, Headline, 1995, p. 239.
5. Paul Davies, *Are We Alone?*, Penguin Books (Orion Productions), 1995. Preface p. xi.

Chapter 3: The Mind's Eye

One touch of nature makes the whole world kin.

William Shakespeare (*Troilus and Cressida*)

The idea that humans can communicate directly from mind to mind is probably as old as civilization itself, and a limited form of this skill may derive from a time before that. During the past hundred years, countless experiments have been conducted in an effort to pin down the phenomenon and to attempt to explain how it could work. These have gradually become more refined and researchers have managed to eliminate almost every possible way in which the phenomenon could be faked, but the scientific community still sees such tests as little more than elaborate tricks.

The reason for this scepticism is that telepathy, if it does exist, is an elusive phenomenon which has been dubbed *jealous* because it often does not work properly in the presence of a sceptic, neither does it usually comply with the wishes of the experimenter. To the non-believers this is a symptom of what is called a nonfalsifiable hypothesis or a hypothesis against which there can be no evidence. An example would be for me to say my pet cat, Sophie is in fact the Creator of the Universe and that the Universe did not exist before she created herself five years ago. You may argue that you remember the world six years ago, but I could then counter-claim that when Sophie made the Universe, she put the idea into your head that you lived before that date and implanted memories and images to accompany it. There is no way such a hypothesis can be disputed using pure logic or reasoning; it is nonfalsifiable.

But with many paranormal phenomena there are other reasons for scepticism. One of the central tenets of science is repeatability. If a scientist claims to have observed a physical phenomenon and conducted experiments to measure the effect, it is only taken seriously by the scientific community if the effect can be repeated under identical conditions by other scientists. If the experiment is unrepeatable, serious doubt is cast upon the original evidence. An example of this is the cold fusion experiments mentioned in Chapter 1. When scientists around the globe followed precise instructions provided by the originators of the theory and could not obtain the same results, they gradually realized that a mistake must have been made in the seminal experiment. The same attitude has rightly governed the approach of the scientific establishment towards telepathy and many other paranormal phenomena. But, for the purpose of seeing how telepathy could work, we have to suspend disbelief and look closely at possible mechanisms.

What do we really mean by telepathy? The image from science fiction is of a being with the power to look into another's mind and to pluck out thoughts as they wish, or to manipulate the thoughts of their subject, to make them do things against their will. This though is an extreme form of telepathy: it may well be that all of us are capable of a type of telepathic experience based not upon supernatural abilities, but enhanced perception.

The psychologist James Alcock illustrates this idea when he says: 'I was standing in a cinema waiting to buy some popcorn, and was idly recalling a conversation I had once had with the brother of a colleague . . . A few moments later I turned around, and there about 30 feet away was the man himself.'[1]

The initial response to such an event would be surprise and a feeling that perhaps you had just had a paranormal experience, but that would involve mixing cause and effect. What had really happened in this case was that Alcock had noticed the colleague's brother subliminally *before* he started thinking about him. This is a well-known phenomenon in psychology, although for a long time, this too was scoffed at by psychologists. The effect is called 'backward masking' and has been validated by laboratory experiments.

If a subject is shown an image for about a tenth of a second, they will be able to recall some of the features of the image, but if they are shown a tenth of a second flash followed by another image lasting longer, the first is forgotten, although it can influence the reported description of the second. For example, if a picture of a man is shown for a tenth of a second followed by a different man holding a knife, the subject's description of the knife-wielding figure is influenced by the characteristics of the first subliminal image.

This is a rare experience for the majority of people – hence the feeling that something supernatural is going on – but other mental skills have been studied that could help to explain a natural process, something that might be called *ultra-sensory perception* or USP.

We are all now familiar with body language or non-verbal communication, but few people use it consciously. Facial expressions, head movements, body positions, tones of voice and even odour can send us subliminal signals, the interpretation of which is usually subconscious. Head and facial movements give the most information about the type of emotion being expressed and we are all instinctively able to interpret these, but politicians and actors are trained to pick up more subtle signals, to utilize an enhanced, but quite natural, subconscious skill which we all possess.

These skills involve just our five senses, but instead of the information being processed by our conscious minds on full-alert for signals such as those we receive when we are concentrating on something, the images are filtered into other regions of the brain and siphoned off without our knowing it. Often they are only interpreted later. And our natural senses can sometimes surprise us with their sensitivity.

Recently, the phenomenon of the cocktail-party syndrome was recounted in magazines and newspapers and caused a flurry of excitement. Cocktail-party syndrome is the buzz phrase used to describe everyone's enhanced receptivity to their own name. Above the general hubbub of a cocktail party or some other noisy environment, we can pick out our name, even if it is whispered on the other side of the room.

The cocktail party syndrome is nothing more than a survival mechanism left over from early human development. If our name is mentioned it means we might be called upon in some way. It may signal the approach of an aggressor or a rival attempting to identify us or it could be for a reason beneficial to us, something we do not want to miss out on. Our ability to subliminally notice things of which we are not consciously aware is the result of a filter system. If we were to give equal importance to everything we picked up with our senses, we could not focus on what is important and get on with our lives. Our brains are programmed to know what is important and what is not and we can grade these sensations.

It is quite possible that modern humans are not as adept at interpreting subliminal messages as our ancestors once were. The reason for this is that we have developed certain skills to a very high degree because of the culture we have established and neglected others, causing them to gradually fall into redundancy.

Our ancestors, some suggest, used hand signals and other forms of body language before verbal communication was developed 50–75 000 years ago. One obvious reason verbal communication was developed and adopted over sign language is that it freed the hands and did not need a visual element. If a hunter was trapped by a wild animal as he returned to the tribe, he could call for help even whilst clinging to the branches of a tree as the animal circled below; sign language did not offer this versatility. Sadly, although we have retained a little of our ability to read body language and to sense a range of signals borne by smell, most human cultures have not been able to evolve a twin path of development, keeping in conscious touch with our primitive instincts whilst adopting the sophistication of language.

Although we do not notice them, these skills remain and utilizing them could produce humans with highly developed talents most of us would consider paranormal. Imagine the skills specially-trained individuals could display if their natural talents and their ability to communicate better with their unconscious facilities were developed to the equivalent of Olympic standard. Such people would be seen as genuine telepaths.

This is what Isaac Asimov had in mind when he created the Second Foundationers in his epic books comprising the 'Foundation' series. Asimov did not believe in thought-transference by any mystical means and had no need for it in his fiction. Instead he imagined a community that had been completely isolated from the rest of humanity and selected for their innate skills. They were then trained by adepts to utilize the full range of their natural human senses and taught to read the slightest movement in others including involuntary muscle movements. They could analyse the meaning of every voice inflexion, could sense odours at long distances and understand their meaning. Not surprisingly, such individuals appeared to the uninitiated as superhuman and their powers viewed as paranormal.

This scenario is perfectly within the bounds of accepted science and involves no use of supernatural powers but simply the development of inherent abilities stretched to the limit. In this way, such a programme would be no different to the process by which Linford Christie or Pele became great sportsmen.

In fact, we do not even need to draw upon such exceptions. Consider the extraordinary sensitivity of the wine-taster's palate or the link between the fingertips and the brain of a blind person reading Braille, or of the musician with perfect pitch.

There is also the possibility that if the human body is placed in an unusual environment where the range of signals is extended, we can utilize aspects of our senses we did not realize we possessed. When astronauts were first sent into space during the early 1960s they reported seeing strange flashing lights. One explanation for this was that they were seeing images that would be beyond those normally experienced on Earth. Their eyes detected the images, but their brains were untrained in translating the signals.

Some relatively simple animals also display skills that appear on the surface to be paranormal. The dogfish catch flatfish by picking up tiny muscle movements of their prey hiding invisible beneath the sand of the seabed and some species of eel possess a 'net' which surrounds them, a form of electromagnetic field that can detect the presence of other creatures within range, a little like radar. Again, there is nothing supernatural about these abilities; they are the result of particular evolutionary paths along which these animals have travelled.

Within human communities, some of these skills are still to be found in a limited sense. The Cuma Indians in the San Blas Islands off the coast of Panama are said to use odour as a way of helping to judge one another's mood, and clasp each other under the arm pits and then sniff their palms when they meet. Our sanitized version is to shake hands, but people from developed nations are also subliminally sensitive to smells and these can influence our feelings towards others without us realizing it.

Everyone has sensors in their joints and muscles which tell us where we are in three dimensions, and others in our inner ears to pass on information about gravity and movement. We also have elaborate systems within our bodies which regulate temperature, monitor the level of chemicals and control our highly sophisticated metabolism. Perhaps there is no need to look beyond these to interpret most telepathic experiences.

The difference between all of these sensory effects and telepathy is a question of scale. All forms of ultra-sensory perception, from differentiating closely associated smells and flavours to the ability to register an image lasting only a tenth of a second, are measurable responses. If telepathy does exist and is an alternative phenomenon (as opposed to a collection of natural abilities), it must operate in one of two ways. Either it uses an extreme form of an information transmission system with which we are already familiar – most probably a region of the electromagnetic spectrum, or else it utilizes a completely unknown form of information transfer. If the former is the case, the reason we have not been able to detect it is because our instruments do not operate within the necessary range or are too insensitive. If the second possibility is true, we may never develop machines with which we could detect or measure the effect, at least not until this alternative means of conveying information is understood. Marconi, the inventor of the radio, would not have thought of constructing his prototype without being aware that radio waves existed. If he had built a radio in ignorance, finding what to him would have been hypothetical radio waves, it would have been a rather hit and miss business.

The human brain contains about ten billion neurones or nerve cells, any one of which may have many thousands of connections to other cells, making it the most complex machine known to humanity. Each neurone acts like a binary gate in a computer, switching on or off, and in this way, thoughts, emotions, decisions and inspirations are formed and transmitted through a vast network. The neurones are linked by axons, the tips of which do not actually touch. Instead a signal is passed along the axon and crosses what is called a synaptic gap to another neurone in the space of about a millisecond (one thousandth of a second). This impulse has an electrical potential of about 120 millivolts and is produced by chemical means, using charged atoms called ions

Figure 3.1

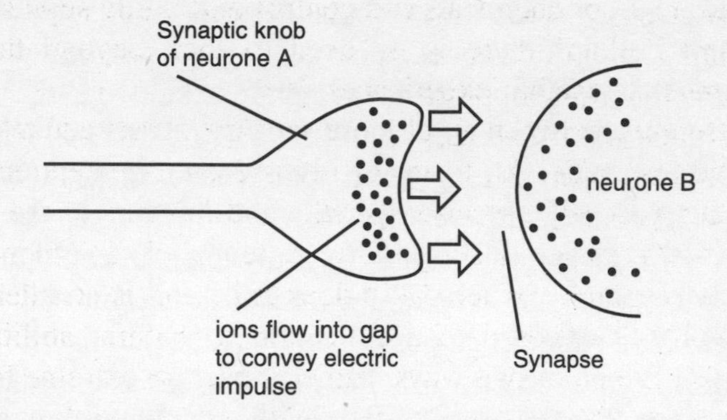

which are triggered to fill the synapse and make the connection to a neighbouring neurone.

It has been suggested that if telepathy is possible, the mechanism by which thoughts are transferred has to be explained at least to this level of the process. This may mean that some form of leakage occurs during the countless individual steps that constitute a thought and that a telepathic individual can somehow pick up this leakage and translate it into meaningful images.

This would be a little like using a phone-tapping device to listen in on someone under surveillance, except for one very important difference – the level of complexity involved when dealing with the human brain is several orders of magnitude greater. Phonetaps work on the elementary principle of siphoning off a signal travelling along a single relatively large wire or using a remote receiver to access a signal sent between two individuals. By this analogy, the telepath (the mental phonetapper) would access a single impulse between two neurones, but this would of course be quite useless because the simplest thought or instruction requires many thousands of neurones working in unison and a single 'neurone tap' would gather almost nothing of use.

Perhaps a telepath can tap into a multitude of neurones simultaneously, but the difficulty with this idea is the decoding process. How would the telepath's receiving equipment manage to decipher all the trillions of impulses racing through the brain at any given moment? They may want to learn what their subject is thinking about a particular subject and all they receive is the interference of signals passing on instructions to release enzymes, to scratch a leg or to control the bladder.

Continuing with our phonetapping analogy, a telepath would be tapping into the most sophisticated telephone exchange imaginable, trying to pluck out a few tens of millions of related conversations simultaneously, then piecing them together to make a coherent message. Of course, different regions of the human brain are responsible for different functions, so if the telepath could tune into certain regions then the task might be a little easier.

An alternative suggestion, postulated by some parapsychologists, is that during the process of impulse transmission, the brain releases hypothetical particles called *psitrons*. Although supporters speculate that these particles would be released in large numbers, they have not yet been detected. According to enthusiasts of the theory, this is because psitrons possess no mass or energy.

Although this sounds ridiculous, the notion of similarly ethereal particles is not without precedent. In the early days of quantum mechanics, physicist and Nobel Laureate Wolfgang Pauli predicted the existence of chargeless, almost massless particles called neutrinos, which were eventually observed in 1956.

The psychologist Carl Jung collaborated with Pauli on a book exploring the paranormal entitled, *Interpretation of Nature and Psyche*[2] and maintained an open-minded approach to the possibility that telepathy could be explained by resorting to the esoteric fringes of known physics. He went as far as to suggest that: '. . . the microphysical world of the atom exhibits certain features whose affinities with the psychic have impressed themselves on physicists. Here, it would seem, is at least a suggestion of how the psychic process could be ''reconstructed'' in another medium, in that, namely of the microphysics of matter.'

But this is a terribly vague statement. It is easy to draw a hypothetical link between two disparate subjects like psychic phenomena and physics in this way, but it does not address the key facts. The crucial difference between the neutrino and the hypothetical psitron is that the former fits perfectly into the family of known particles; it plays a recognizable role and was predicted by the strict mathematics of quantum mechanics before it was detected. Anyone can think up an imaginary particle to explain a phenomenon, give it an appropriate name and suggest that it lies at the root of an unprovable process. Believers have even gone so far as to suggest that psychic powers do not work in the presence of sceptics because the wills of the doubters suppress the action of these particles. This is a perfect example of a nonfalsifiable hypothesis at work.

Psitrons have never been detected but this does not mean they do not exist. It is possible that some form of field or resonance or even an array of particles are produced as a by-product of brain activity, but until these are found and their properties understood they should be considered as pure conjecture.

The brain does of course produce measurable potentials which are associated with different brain states, and their discovery by Richard Caton in 1874 raised hopes that science had stumbled upon the method by which telepathy worked. The reality is sadly much more mundane.

Four distinct types of rhythm have been identified in the human brain and these correspond to different brain states. The waves are due to electrical activity and manifest as oscillating electrical currents.* They are detected by a machine called an electroencephalograph (an EEG) which picks up the tiny electrical impulses by attaching electrodes to the scalp and amplifying them. The signals can then be read out on a graph.

The four distinct brain waves are placed in frequency bands and measured in cycles per second or Hertz (Hz). When the brain is resting and relaxed it produces alpha rhythms which are detected between 8 and 14 Hz. Beta rhythms correspond to activity and predominate when the brain is working, solving problems or facilitating movement such as walking, or running. These rhythms occur between 13 and 30 Hz. At the other end of the brainwave spectrum are delta waves which are produced during sleep. These have widely spaced peaks and oscillate at between 1 and 4 Hz. Finally, theta rhythms, which are produced when the brain is in a deep sleep or a trance state, resonate between 4 and 7 Hz.

* These brain rhythms should not be confused with the idea of biorhythms which are thought by enthusiasts to be quite different in origin. See Chapter 7.

Brainwaves captured the public imagination during the 1970s and the market was flooded with devices that were claimed to be capable of inducing alpha rhythms instantly. This was actually utilizing what yogis and Zen adepts had known for a long time, that individuals are able to control their own brainwaves. It led to an awareness of what is now thought to be a fourth state of consciousness, a deep relaxation state or meditative condition corresponding to theta rhythm production.

Although research into brainwaves has generated benefits for medicine, it has not led to the source of telepathic ability. EEGs are used extensively in psychiatric treatment and are particularly useful in the treatment of epileptics who exhibit disrupted brainwave patterns. The electrical impulses detected in the cerebral cortex (the outermost layer of the brain, a few millimetres thick) reflect the overall brain state; they cannot be deconstructed in order to draw off particular thoughts or even emotions.

Yet some researchers claim that certain rhythms are more pronounced when subjects are believed to be acting telepathically. These results are based upon the use of an EEG machine during telepathy tests and show that alpha rhythms accompany supposed thought transference. But this is misleading because alpha waves are most noticeable when an individual is in a relaxed state, which

Figure 3.2

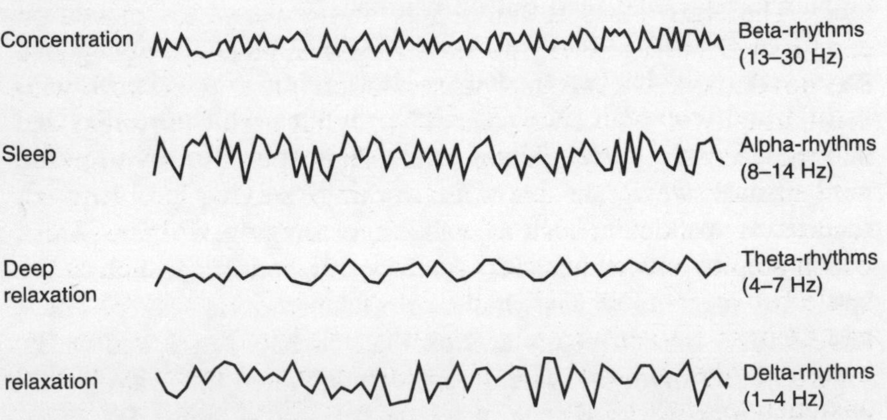

is also the brain state most clearly associated with telepathic successes in laboratory tests.

Putting aside attempts to find particles or wave-forms to explain telepathy, some parapsychologists have suggested that the telepathic experience is an holistic effect, some form of response to a network made up of all human consciousness. It may have been this concept that one of the founders of quantum theory, Erwin Schrödinger had in mind when he said: 'I – I in the widest meaning of the word, that is to say, every conscious mind that has ever said or felt "I" – am the person, if any, who controls the "motion of atoms" – according to the Laws of Nature.'

The way in which a 'human network', or a network that includes all living beings could operate, is little understood. Some researchers have made an attempt to clarify the concept, or to link it with aspects of biology and psychology, but the results have drawn only further controversy and in some cases, confusion.

Jung postulated the idea that there were two forms of unconscious awareness, personal unconsciousness and what he dubbed the 'collective unconscious'.[3] This he saw as an inherited set of images common to all human beings. He called these images 'archetypes' and believed each to be symbolic of a deeper aspect of human imagination including ideas such as having parents, having children and death.

In the hands of master manipulators, archetypes are powerful tools. The Nazi rallies at Nuremberg during the 1930s played on deep-rooted human fears and hopes by manipulating archetypes, and novelists, musicians and painters often imbue their work with them, usually without conscious effort. It has even been suggested that the success of any form of art depends upon the artist linking with an audience via the use of archetypes. An example is the huge success of Tolkien's *Lord of the Rings* or the *Star Wars* films, which employ archetypes throughout, symbolic images such as the battle between 'good' and 'evil' and the 'wise old man' (Gandalf and Obiwan Kenobi). In the same way, the *X-Files* often uses the image of 'the stranger' and plays on universal human fears and anxieties to great effect.

A related phenomenon is a process called 'formative causation', which was first postulated and popularized by the British biologist Rupert Sheldrake in his book, *A New Science of Life*, published in

1981.[4] In essence, Sheldrake suggests that systems 'learn' or that it is easier to repeat something if it has already been done. The mechanism for this is called 'morphic resonance'.

Initially, this sounds like a rather vague notion and Sheldrake has indeed been savaged by orthodox scientists around the world from a variety of disciplines, but, he has spent the past fifteen years conducting experiments which he claims verify the concept repeatedly.

One of Sheldrake's demonstrations of the principle is based upon linguistic patterns. He asked a Japanese poet to send him three similar verses. One was a meaningless string of words, the second a freshly composed verse and the third, a well-known poem familiar to Japanese school children. He then showed the three pieces of writing to a group of westerners, none of whom could speak any Japanese. What he discovered was that all of the subjects found it far easier to memorize the traditional poem than the other two.

His conclusion was that the traditional poem had somehow become ingrained into human consciousness via morphic resonance. Based upon this and numerous other tests, some involving living beings as well as experiments involving inanimate matter such as growing crystals, Sheldrake and his supporters believe that all things resonate with their own kind – 'like resonates with like'. In other words, there is a network of human interaction, and similar 'fields' around other species, other inanimate objects which influence their behaviour. But it could be argued from the opposite perspective. Using the example of the Japanese poem; could it not have been that it had survived and become familiar to school children because it was easy to learn? Could not certain functions become easier with repetition because those doing them have a natural affinity towards them and avoid tasks that do not come naturally?

If morphic resonance or the collective unconscious are realities, they may show us alternative ways in which minds may communicate. The traditional image of telepathy is that it occurs via pseudo-physical means, facilitated perhaps by rays or particles, but the truth may be far more subtle. In a sense, any artist may be communicating with their public telepathically by tuning into

Figure 3.3 Zener cards

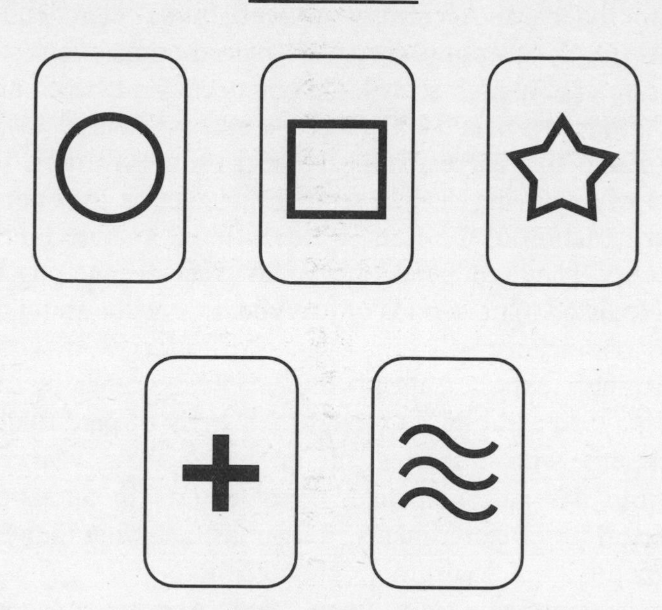

archetypes or using morphic resonance to cross the barriers of space and time. Perhaps there are special people (and other animals) that have a greater sensitivity towards these resonances, are more adept at manipulating archetypes in far more sophisticated ways than we usually experience.

Whatever mechanism lies at the root of the telepathic experience, parapsychologists have been obliged to follow the traditions of science in attempting to demonstrate psychic phenomena. This is really the only way in which they can hope to convince a sceptical scientific community and to develop an understanding of what is happening, if anything.

Researches into paranormal phenomena began during the nineteenth century, but it really came of age with the American parapsychologist Joseph Rhine who summarized his findings in his 1934 book, *Extrasensory Perception.*[5] Rhine worked at Duke University in North Carolina and pioneered the use of what became known as Zener cards (after their inventor Karl Zener). There are five designs on the cards, a circle, square, star, plus sign and three wavy lines.

These experiments involved the experimenter taking the top card from the pack and attempting to transmit the information on the card to the 'reader' or subject to see if they could identify the symbol on the card. According to probability, there is a 20% chance of simply guessing correctly, but in some trials subjects obtained remarkably high scores. On one occasion a participant got 588 'hits' from just over 1800 trials – a success rate of 32%. This does not sound like a great improvement on the average, but the likelihood of achieving such a score by chance is astronomically high. In a variation of the tests, Rhine offered a subject $100 for every success. They produced a run of 25 successes, netting $2500, a result calculated to have odds of just under three thousand million million to one against.

After Rhine's experiments hit the headlines, other researchers followed his lead and rapidly brought the study of parapsychology into disrepute with a series of infamous fakes. Since then parapsychologists have expended great efforts in attempting to develop fraud-proof experiments to demonstrate what they believe to be a genuine and measurable phenomenon.

Modern experiments rely upon random number generators which churn out numbers which are supposed to be completely without pattern, a little like the lottery machines wheeled out each Saturday evening. These experiments are seen as more reliable than the early Zener card tests, but the objections of the sceptics, the self-styled *psi cops* have driven the researchers to new heights of sophistication.

The latest experiments are known as *Ganzfeld* or blank-field studies. These involve participants undergoing sensory deprivation in a form of isolation tank. Ping-pong balls are placed over their eyes and white noise is played through headphones. The experience has been compared to staring into a formless fog. After about fifteen minutes most subjects experience hypnagogic images, those often experienced on the edge of sleep. A sender, usually a friend or close relative of the subject, is placed in an acoustically shielded room from where they try to send an image, usually a one-minute long video sequence or a static image.

This research is being conducted at a number of centres around the world including a site at the University of Edinburgh founded in 1985 with a bequest from the Nobel-prizewinning author, Arthur Koestler, who was a great believer in paranormal phenomena. So far, there has been little in the way of conclusive evidence to support traditional ideas of telepathy coming out of these experiments. Like Rhine during the 1930s, the teams have found rare individuals who have achieved impressive scores which lie far outside the normal range of probability. Unfortunately, these results are usually unrepeatable and therefore cannot be deemed in any sense conclusive by orthodox science.

One of the most striking implications from the vast number of experiments conducted during this century is the idea that telepathic ability can apparently be enhanced by a wide variety of factors. The example of the man who produced twenty-five hits in a row when given the incentive of financial reward is a mundane example. Experimenters have become interested in the idea that if other senses are suppressed then telepathic powers can come through more readily. This is the reason for isolating the subject in the *Ganzfeld* experiments but it has also formed the basis of experiments linking sleep with telepathy. Once again the results show that a small number of impressive, but unrepeatable events suggest more pronounced telepathic ability if the other senses are dampened. One researcher has correctly pointed out that: 'If psychic powers exist, everyday experience shows that they must be very weak for most people.'[6]

Other anomalies could have an influence upon telepathic power. It has been found that children with mental defects score higher in telepathy tests and in one set of experiments, a child known simply as 'the Cambridge boy', who had been born with physical and mental disabilities, achieved well above average scores when his mother was present.[7]

The explanation for this is that if telepathic abilities really do exist then they may be more useful to individuals who cannot communicate in the conventional manner or have their other senses suppressed in some way. There have also been a number of unconfirmed cases of telepathic powers becoming apparent during life-threatening situations. These have been dubbed by parapsychologists as 'need-determined' cases or 'crisis telepathy', but dismissed by orthodox science as apocryphal.

According to some researchers, this form of telepathy could be explained by its survival value and might even be a genetically favoured trait. Humans, they argue, have submerged this talent with other more readily developed and utilized abilities, but some rare individuals are more in tune with this power and it comes to the surface in an emergency.

It has also been supposed that other species display this facility. During the 1970s Soviet parapsychologists attempted to demonstrate this effect experimentally. They took a set of newly born rabbits away from their mother and killed them at set, recorded times. The mother was wired up to an EEG and her brain patterns monitored. According to official reports, the mother rabbit displayed sharp electrical responses at the precise moment each of her offspring were killed. Unfortunately, because news of this experiment leaked out from Soviet Russia, it is difficult to verify and nobody in the West has so far re-investigated it.

Sceptics continue to pour cold water on the entire phenomenon of telepathy. One of the most usual arguments is to ask why telepathic individuals do not use their skill to win lotteries or to chalk up staggering success at the race track. They also wonder why in lab tests the talents of the claimants mysteriously vanish.

The problem with telepathy is that a century of investigation has turned up little evidence that complies with standard scientific practice. But, during this same period, science has performed quite apparent wonders, from curing diseases to placing humans on the surface of the moon.

A study by the highly sceptical National Research Council in the United States found in 1988 that there were what they called 'problematic anomalies' in some experiments that could not be explained; in other words, incidents of success that could not be accounted for merely by chance. And, despite the apparent lack of evidence, many people believe in telepathy. In one survey, 67% of people questioned said they had experienced ESP.

Psychologists have noticed that 'a sense of deep personal conviction may be the key to achieving good results,'[8] but this is surely not the whole story. There is still no satisfactory scientific explanation for what supporters claim has occurred during a growing number of experiments, but it would be unscientific to conclude from this that telepathy is imaginary. It may just be that we don't know how it works.

References

1. James Alcock, *Parapsychology: Science or Magic?*, Oxford, 1985, p. 86.
2. Carl Jung and Wolfgang Pauli, *The Interpretation of Nature and the Psyche*, New York: Pantheon, 1955.
3. See Carl Jung, *Man and His Symbols*, Aldus Books, London, 1964.
4. Rupert Sheldrake, *A New Science of Life*, Blond and Briggs, London, 1981.
5. J.B. Rhine, *Extrasensory Perception*, Boston: Bruce Humphries, 1934.
6. Chuck Honorton, quoted in: 'Roll Up For The Telepathy Test', *New Scientist*, 15 May 1993, pp. 29–33.
7. See Terry White, *The Sceptical Occultist*, Arrow, 1988, Chapter 2.
8. Susan Blackmore, quoted in: Terence Hines, *Pseudoscience and the Paranormal*, Prometheus Books, New York, 1988, p. 82.

Chapter 4: Moving Heaven And Earth

O the mind. mind has mountains: cliffs of fall
Frightful. sheer. no-man-fathomed.

Gerard Manley Hopkins

Telepathy is really only the tip of the psi powers iceberg. If we think of telepathy as being 'mind communicating with mind', then what are the chances of mind interacting directly with the physical world, with matter?

Psychokinesis or PK is defined as: 'The apparent ability of humans to influence other people, events or objects by the application of will, without the involvement of any known physical forces.'[1] Whereas telepathy could be described as an interaction between two psychic 'fields' or 'forces', PK involves an interaction between the mental and the material, so it is one stage further down the psychic route. To the sceptical, this means a marriage from hell, but from the perspective of parapsychologists it is merely a natural progression from the more prosaic telepathy.

PK has been the subject of even more experimental work than thought-transference, and according to how you view this evidence, it has either been proven beyond doubt to be a genuine, natural process or else all the tests and experiments conducted during a period of almost a century have been faked or may be explained by other factors.

The earliest serious attempt at trying to quantify the concept of PK is attributed to the parapsychologist J.B. Rhine who worked on telekinesis experiments concurrently with his attempts to pin down telepathy during the early 1930s. He was led into the subject when a young gambler told him he could influence the fall of a die by will power alone. Rhine immediately set about conducting thousands of tests on scores of subjects, in an attempt to reach a statistically meaningful conclusion which would show if there was any such effect.

He based the test upon participants trying to will two dice to produce a score of more than seven. As with the results of his telepathy tests, Rhine and others found that most people achieved scores that deviated little from the values expected by chance, but once in a while he turned up individuals whose scores did not fit the normal pattern. On a few occasions, he found someone whose scores deviated from the average to such a degree the probability of it happening by accident was sometimes placed at millions to one against.

In one set of experiments Rhine conducted 6744 tests. These should have produced 2810 successes by chance alone, but one subject achieved a score of 3110, a deviation calculated to occur by chance only once in a billion tests.

Rhine attracted criticism of his experimental methods almost immediately, but, as with his telepathy experiments, he went to great lengths to debug his tests from any chance of trickery or unintentional influence. After nearly thirty years of such experiments, he reached the conclusion that 'the mind does have a force that can affect physical matter directly.'[2]

Since Rhine's seminal investigations, literally hundreds of groups around the world have conducted other forms of PK tests. During the 1970s researchers lead by Helmut Schmidt at the Mind Science Foundation in San Antonio, Texas replaced Rhine's die with a Geiger counter. The reason behind this was that Geiger-counter readings derive solely from radioactivity which is produced by the breakdown, or 'decay' of radioactive nuclei. This decay process is completely random and as close to fraud-proof as parapsychologists could hope to get.

Schmidt asked participants to attempt to influence the read-out from a Geiger counter – usually a display showing series of flashing lights or an oscilloscope screen showing a wave pattern. Variants on this theme were developed in the 1980s using white noise patterns generated electronically.

One of the leading parapsychologists currently working with PK is Robert Jahn, an engineering professor based at Princeton University in New Jersey. He developed the white noise experiments and has gone on to try different versions of what he calls random event generator – a machine that produces random displays or number sequences – the electronic equivalent of tossing a coin thousands of times. The generator is fitted with a collection of safety devices to detect any change in temperature, the influence of external magnetic fields or any physical disturbance such as tilting or weighting the machine in any way.

Ignoring the cynicism and sometimes open hostility of his orthodox colleagues, Jahn has spent the past 15 years conducting millions of tests, using over 100 subjects to see if there is any deviation from chance. What he has found is still inconclusive. Taking the collection of experiments and subjecting them to statistical analysis, he has found that there is some effect which he judges to be about a 0.1% deviation from pure chance. In other words, on average, one thousand tests throws up one significant deviation from what was expected.

If this all sounds unconvincing, there are other more worrying aspects to Jahn's experiments. He conducted a large number of trials where the subject was placed far away from the random event generator in his laboratory. Some of the subjects were asked to attempt the test from as far away as Africa and England, but confusingly, Jahn found that their success rate did not vary at all with distance.

In another set of experiments, he asked his subjects to make their attempts at psychokinetic influence up to several days before the test and again found that there was no difference in the quality of the results. Others have discovered the same odd anomaly. In a different collection of tests conducted by Helmut Schmidt, he disconnected the random event generator and substituted the 'live' read-out with a recording of the signals from the previous day, but didn't tell his subjects what he had done. He found that, if anything, the results were better than they had been in the usual experiments.

PK enthusiasts have produced a very odd explanation for this. The subjects, they claimed, were practising what has quickly been dubbed *retroactive psychokinesis*. This involved them sending thoughts back in time and space to the previous day and affecting the read-out.

Astonishingly, many parapsychologists subscribe to this explanation. Cynics merely call it a nonsensical explanation for an obvious and telling flaw in the entire theory and practice of parapsychology. Yet there is actually another far more mundane explanation which appears to have escaped both camps. If the results on Day 2 are affected by the subject even though the display from Day 1 was being shown to them, perhaps their minds were interacting with the playback device. If we are trying to demonstrate PK, there is no reason why the subject might not control some machine feeding the fake display any more than they might alter the pattern from a random event generator. After all, the playback device is certain to be some form of tape recorder or digital device and as susceptible to PK as any other material system.

So, putting to one side the views of the extremists from either camp – the ardent sceptics and the whole-hearted believers, what scientific conclusions can be drawn from the vast range of PK experiments?

Enthusiasts point to the rare outstanding anomalies and conclude that something odd is happening and that this must point to proof of PK. But they seem unperturbed by the fact that those cases which deviate greatly from chance are quite exceptional and very rare. More common is a tiny perturbation from the expected which may perhaps point to a weak force at work, or it could be due to a number of other anomalies.

One such anomaly could be stray magnetic fields or electric currents. It has been found recently that some people who live near power cables sometimes experience physical illness and depression. This is thought to be due to the close proximity of the brain to powerful electrical currents. All electrical impulses have associated magnetic fields and those around carriers such as national grid power cables interact in some way with the similar but far weaker magnetic fields created by the electrical signals produced in the brain. Via some unknown mechanism, these disturbances manifest as physical and emotional instability. In a similar way, it is possible that magnetic fields some distance from the test centre could have a weak effect on the machinery used for the tests. Another source might be electrical interference by leakage from equipment elsewhere in the lab or even beyond the building. Experimenters have tried to negate this anomaly by encasing the test equipment and the subject in a special container called a Faraday cage which shields them from electromagnetic disturbances.

Such a tiny effect as that observed by parapsychologists could be produced by other natural sources. We all live on a giant magnet. Like most planets, the Earth has its own magnetic field which fluctuates naturally due to movements hundreds of miles beneath the crust. It is also disturbed by fluctuations outside the atmosphere. Sunspots, which are cooler regions on the surface of the Sun, are able to disrupt the Sun's powerful magnetic field and this in turn can alter the Earth's associated field. Such magnetic disturbances might conceivably affect electronic machinery and indeed the fields each of us produces by electrical activity in our brains and from each nerve impulse passing constantly through our bodies.

Other factors to consider are currents of warm or cool air and minute geological disturbances such as micro-earthquakes. Although these factors are almost undetectable, they could be powerful enough to disturb PK experiments.

These objections may sound pedantic, but if parapsychology is to be taken seriously by scientists, it has to play by the same rules as orthodox science. Many critics of PK and other paranormal phenomena cite undue care or unprofessional attitudes to research as the most likely source for their apparently impressive results.

The only reasonable conclusion to be drawn from the millions of PK experiments conducted since the 1930s is that if the results do demonstrate a genuine effect produced by the human brain, then the effect is very small. And because it is so small it is incredibly difficult to measure. Parapsychologists have dubbed this effect *micro-PK* and there is a growing body of evidence to substantiate an anomaly of this type.

In the mid-1970s a psychologist Gene Glass came up with a revolutionary approach to the study of experimental results from parapsychology experiments. He realized the problem with PK was that the effect was so small, usually in the region of 0.1% over chance, that a large number of results would be needed to show up the anomaly created by any genuine paranormal activity. Furthermore, the smaller the effect, the more results would be needed. He called his system *meta-analysis*.

An analogy would be the effort put into tuning a radio. If a signal is strong, such as say a BBC transmission picked up in Southern England, it would be easy to tune into it. If, on the other hand, you were trying to pick up a weak signal from a pirate or independent broadcaster, you might have to spend some time fine-tuning the radio to detect the signal. This fine-tuning is equivalent to conducting a large number of samples or tests.

So Glass's idea was to somehow pool the data from all the tests that had been conducted over a long period. The problem was that the tests carried out since the 1930s were quite different from one another. Some experimenters had investigated the possible effects of PK on falling die or wood blocks, others followed the altering of light displays or the random decay of unstable isotopes using a Geiger counter. But Glass found eventually that using suitable mathematics, results from disparate tests can be merged, which means that the parapsychologist has access to a far larger sample – tens of millions of results taken by scores of experimenters over a period of some sixty years.

One of the best examples of applying meta-analysis to PK experiments comes from the work of a psychologist, Dean Radin at Princeton University's Psychology Department and Roger Nelson, a member of the Princeton Engineering Anomalies Research programme (PEAR). They did not conduct their own experiments but instead tracked down over 150 reports summarizing almost 600 separate studies and a further 235 control studies by 68 different investigators each of whom had been researching the influence of consciousness on microelectronic systems – experiments where the subject was asked to disturb the workings of an electronic random event generator.

To their amazement, they found the probability of the net result deviating from the normal pattern merely by chance was 1 in 10^{35} (1 with 35 zeros after it).

Again, this result does not say for certain that PK is a genuine phenomenon, but it does suggest there is some factor or combination of factors that alters the behaviour of matter other than by the visible, conventional means. Whether this is the effect of human consciousness or sun spot activity, micro-earthquakes or thermal currents is another matter.

Those convinced that the aberrations shown up by meta-analysis are of human origin believe the effect is generated by micro-PK and that this phenomenon is with us all the time. They suggest that many incidents we might think of as coincidence are a result of this force. When we drop a book and it opens at exactly the page we wanted; when we look through a filing cabinet and put our fingers straight on the piece of paper we've been looking for; when we pot the black without even looking. These, according to the supporters of micro-PK, are all examples of a subconscious ability to influence the way matter behaves. Furthermore, it is thought that this phenomenon works best when the subject is not deliberately trying to make it work, when the individual is concentrating on something quite different.

We have all experienced moments when things have either gone extremely well for us or extremely badly – good days and bad days. We've all experienced beginner's luck, moments when you can do no wrong. It is on these occasions, the believers say, that micro-PK is working at its best, and these are exactly the occasions when we are the least likely to be trying to make it work.

A psychologist called Rex Stanford, working at St Johns University in New York State on the east coast of America, has conducted an interesting variation on the usual PK experiment to illustrate the effects of micro-PK. He places his subjects in a locked room and gets them to perform a series of very dull tasks. In the next room is a random number generator. What he does not tell the subjects is that they can only be released from their task and allowed to leave the room when the generator produces a sequence of numbers that appear only once every two or three days under normal circumstances. Yet, on several occasions, some subjects have managed to get out of the room within 45 minutes.

The problem with accepting micro-PK is that the force producing the effect is the same as that involved in macro-PK. Any system which allows us to subconsciously control the way a book lands or the movement of a ball in a game of billiards, football or cricket is the same phenomenon that could allow us to move objects at will. They are on the same scale and would presumably operate by the same wave-form, particle stream or other inexplicable force. And it is not whether these effects occur occasionally or frequently that bothers the scientist, it is how they could occur even once. Because at the root of the dilemma remains the fact that no form of psychic force has been detected, yet we are asked to accept that the mind can interact with matter – the marriage from hell I mentioned at the start of this chapter.

To illustrate the problem, let us consider the physics of PK.

What are the energies involved and is there any compatibility between what is needed and what could reasonably be generated by the brain?

Let us imagine an experiment in which a subject who claims the ability to perform psychokinesis is asked to move an object along a table using just the power of their mind. Suppose our object weighs 100 grams, the mass of a spoon or a pair of spectacles. Now assume the participant is to accelerate the object to the modest velocity of 10 cm per second and to maintain this velocity for a few seconds. If we add a small contribution from friction, the energy needed to do this comes to approximately 1×10^{-4} Joules (one ten thousandth of a joule).

This is a relatively small amount of energy, roughly equivalent to that stored in one millionth of a gram of sugar. But equally, to produce even this much energy from a force which has so far remained undetected by any conventional means requires a conscientious suspension of disbelief. Consider the figures.

As we saw in the last chapter, the brain has associated electric and magnetic fields. Now, we could suppose that the electrical impulses from the action of neurones is responsible for creating the force with which the object is moved. But this field must interact very weakly with the material world, because we cannot pick up the force or any form of tangible interaction with any known instrument. But, let us be liberal and say that one thousandth of the power of the electrical impulse penetrates the skull and reaches across space to the object and accelerates it to a speed of 10 cm per second for a short period of say 3 seconds. The voltage produced in the neurones is approximately 100 millivolts, so this would mean that the human brain would have to produce a current in excess of 0.25 amps to provide the necessary energy. To put this into perspective, a current of little more than half of this (0.15A) passed through the heart would kill a person.

A less spectacular, but equally intriguing form of PK involves a mind-matter interaction of a different type. This is a process known as thoughtography – the art of generating images on photographic paper or on film without the use of chemicals. A famous example of this talent was investigated by the parapsychologist, Jule Eisenbud, during the 1950s. Her subject was a man named Ted Serios who could produce spontaneous images on film merely by looking at a camera. He became a celebrity for a time and produced a book called *The World of Ted Serios*. Although he was never caught cheating, he later attracted suspicion when he was found to be holding a tiny device in his hand whenever he performed his trick. He refused to let anybody analyse the devise, but could not make thoughtographs without it, claiming that it focused his energies.

This form of PK is another example of the power of the mind apparently altering the physical world. This time, some form of electromagnetism may have been able to change the chemical structure of the photographic paper or the materials coated on the film. Under normal circumstances, light of particular wavelengths would impart sufficient energy to the chemicals on the paper or the film to break inter-atomic bonds and stimulate the creation of new chemical arrangements which would collectively produce a photograph. If Serios, or any other thoughtographer, was genuinely creating an image, it must have been through a similar mechanism. Perhaps it is possible for the very weak fields created by the brain to be amplified and focused at a precise set of wavelengths that in some way duplicate the action of light.

So far in this chapter I have concentrated on what might be termed 'peripheral PK effects' – the results of micro-PK. However, in the early 1970s, the world awoke to what appeared to be an example of full-blown psychokinesis and it was captured on film and broadcast to the world on television. Uri Geller had arrived.

Geller's contribution to the paranormal evokes strong feelings both for and against. To some he is a talented psychic, to others he is a little more than a showman, a trickster masquerading as a mystic. During almost twenty-five years in the public arena he has reworked his career many times but has done little in recent years to top the splash of publicity and sensationalism that surrounded his first appearance in Britain in 1973.

At that time a number of scientists took Uri Geller seriously enough to expend effort and resources studying his claims and trying to reproduce his stunts under laboratory conditions. John Taylor, Professor of Mathematics at King's College, University of London overcame the ridicule of some of his colleagues and the resistance of the college authorities in trying to quantify Geller's apparent skills. Although he was initially intrigued by Geller's performance under test conditions, Taylor soon found he could not reconcile the evidence with electromagnetism and was forced to either dismiss the phenomenon or accept that the bedrock of science was wrong. He chose the former.

'When science faces up to the supernatural,' he concluded, 'it is a case of "electromagnetism or bust". Thus we have to look in detail at the various paranormal phenomena to see if electromagnetism can be used to explain them.'[3]

Within a few months of starting the tests at King's, Taylor had switched from being open-minded to denouncing Geller as a conjurer. In his 1976 book *Superminds*, he wrote:

> Uri Geller appears to have posed a serious challenge for modern scientists. Either a satisfactory explanation must be given for his phenomenon within the framework of accepted scientific knowledge, or science will be found seriously wanting. Since such an explanation appears to some to be impossible, either now or in the future, they argue that the Geller phenomenon is incompatible with scientific truth, and that the value of reason and the scientific point of view is therefore an illusion. Will the gates of unreason then be allowed open and drown us in a world inhabited by aetheric bodies, extra-terrestrial visitors, spirits of the dead and the like? Will reason then wholly give way to superstition?

To be fair to Uri Geller, some of his stunts have been impressive. He has been tested by a dozen different research groups around the world, beginning with a team at Stanford Research Institute in California in 1972 before he became an international celebrity. Since then he has travelled around the world and subjected himself to analysis in laboratories in Japan, France, and several in Britain and the United States. He has been video-taped materializing objects apparently from thin air, stopping machinery and of course distorting a vast assortment of cutlery not only on film but before large audiences on live television shows.

Geller himself conveys the impression that he does not fully understand from what source his talents spring and appears to be totally convinced that what he is doing is paranormal. In the final analysis, he may be the only person who will ever know for sure, because if he is a fraud, he will have no intention of ever letting on.

The researcher Arthur Ellison has said of him:

> For some strange reason Geller has a different image of truth which doesn't quite agree about the strengths of materials taught in university. He believes that if he gently strokes a spoon between finger and thumb without putting a terrific pressure on it then it will bend. And sometimes it does ... I've seen Geller take a Yale key ... he stroked it and the end came off ... Geller's image of truth just doesn't involve Yale keys being forever rigid.[4]

In one test, a 1-gram weight was placed on a sensitive scale and the apparatus covered with a glass belljar. The weight on the scale registered as a line on a chart recorder and Geller was asked to alter the reading paranormally. During a test lasting half an hour, he was able to deflect the chart recorder twice, each for one fifth of a second. The first deflection was equivalent to increasing the weight to 1.5 grams and the second to 2 grams.

On another occasion, he was able to dramatically alter the read-out of a Geiger counter which had also been sealed in a unit some distance away across the lab. Then, when he was allowed to pick up the device, he concentrated hard and the 'beep beep' accelerated to a wail which stopped the moment he put down the unit.

In one sense at least, Geller is an accomplished showman. This has been part of the reason for his success. He does not merely produce impressive tricks, but is happy to employ gimmicks such as spoon-bending which so caught the public's imagination in the early 1970s. When he first appeared on British television and bent an assortment of spoons before the cameras, he became an overnight sensation and people throughout the country and from all age groups were rushing for the kitchen drawer as soon as the show was over.

Even today, more than two decades after his first shows, scientists have still not been able to show how he creates the effect. The magician, James Randi, a vehement critic of Geller, has declared publicly that spoon-bending is nothing more than a vaudeville trick and has himself duplicated Geller's performance quite convincingly by sleight-of-hand. Others have taken a more analytical approach, studying the objects at the heart of the controversy – the spoons themselves.

A team in France have taken electron microscope photographs of spoons before and after bending experiments. They tested the crystalline structure of the metal and measured the weight and dimensions before handing them to spoon-benders and again after the experiment. These tests have revealed little except that there appears to be a local hardening around the area of the distortion said to be similar to that caused by a considerable pressure applied externally and concentrated on a small region. Clearly, no observable external pressure had been used, so the experimenters were forced to conclude it had been produced 'internally'.[5]

Other tests have shown that as the bending occurred, the metal of the spoon takes on the consistency of chewing-gum in the region of the distortion. Interestingly, this effect can be caused by using corrosive chemicals but it always results in weight loss and visible corrosion of the metal surface, neither of which has been observed in these PK tests.

So what are the possible mechanics of spoon-bending? If it is not a conjurer's trick, how could it be done?

As John Taylor points out, scientific explanation is limited; it's a question of electromagnetism or bust. So if Geller and other spoon-benders are able to focus electromagnetic waves into the spoon what sort of energy would they need to produce?

Metals are substances in which the atoms are arranged in a uniform lattice. The shape and characteristics of this lattice depend on the type of metal. If we consider for simplicity's sake that a spoon is made of iron, we need to look at the way in which the lattice of iron could be distorted by PK or any other method.* To do this we will need to relate the energy requirement to that used to distort the metal more normally – by the use of heat.

Iron melts at 1808K (1535 °C), but of course we don't want to actually melt our spoon, just soften it enough to distort its shape. A better value to consider is something called the latent heat of fusion of the metal which is the energy needed to convert solid iron into liquid iron – to disrupt the metal lattice. If we imagine a spoon weighing in the region of 100 grams, perhaps only 20% of its mass is involved in the distortion, so we only need to consider the energy needed to alter 20 grams of iron.

Obviously we do not need all the energy required to convert iron

* They are more typically made from stainless steel which is in fact much tougher and resistant to heat and pressure than pure iron.

from a solid to a liquid, because all we wish to achieve is the 'chewing-gum' consistency observed by some experimenters. Let's be conservative and say that the energy needed to create this state is only 10% of that required to melt the spoon. So, the energy needed is equivalent to 10% of the latent heat of fusion of 20 grams of iron, a figure which turns out to be around 600 joules.

Earlier we saw that the energy needed to move a 100 gram object along a smooth surface was a tiny fraction of this (one ten thousandth of a joule). The electrical energy Geller would need to generate in order to provide 600 joules using only the electrical potential of neurones in his brain, would require a current in the region of 150 amps – the sort of currents used to power heavy-duty electronic machinery.

The empirical analysis of spoon-bending, and indeed all PK activity, does not paint a rosy picture, but then that should be expected – it truly is a marriage from hell. Using the forces we know and understand, matter cannot be manipulated by the energies associated with thought or at least the outward, measurable manifestations of thought – electrical currents in the brain. So we are left with explanations that lie outside the usual scope of the physical and natural sciences.

At this point parapsychologists often resort to drawing in ideas from the fringes of physics such as quantum mechanics, but these do not offer a satisfactory explanation and are often used willy-nilly. Because quantum mechanics is itself a weird area of physics, it is also vulnerable.*

One of the most telling aspects of PK research is the fact that, like telepathy, psychokinetic effects have no respect for distance. The set of experiments conducted by Robert Jahn at Princeton University in which subjects were apparently able to alter read-outs up to several thousand kilometres from the laboratory with as much ease as when they sat a few metres from the detector illustrates this point.

Almost all known physical forces operate according to the inverse square law. What this means is that, all other factors being equal, as distance is increased, the effects of force diminish by the

* I'm of the opinion that quantum mechanics has little to do with PK or telepathy but may be argued as an explanation for the phenomenon of prediction and synchronicity which is discussed in detail in Chapter 6.

square of that distance. For example, Isaac Newton described in his masterwork *Principia Mathematica* published in 1687 how the force of gravity operated via the inverse square law.

By Newton's reasoning, if planet A circles the Sun at a given distance, it will experience a certain gravitational attraction towards the Sun. If planet B with identical mass orbited at double this distance, it would experience only a quarter of the gravitational attraction experienced by planet A, (the inverse of 2 squared). An identical planet C, orbiting three times as far away as planet A would feel only one ninth the gravitational attraction experienced by A, (the inverse of three squared).

All known forces operate by this inverse square law, but, supporters of PK argue that there are other forms of energy transmission which do not. The most popular argument is that the intensity of radio signals does not weaken very much with distance. From this enthusiasts draw an analogy between radio and 'thought waves'.

The first part of this statement is true. If radio signals are generated by a powerful enough transmitter and fine-tuned using what is called a signal-optimizing system, their intensity does not diminish to any large extent over reasonable distances. But, this is not really the point. Radio signals convey information, they do not move objects or enable the bending of spoons; some form of force is needed to do these things. Furthermore, if PK operates by electromagnetic radiation of which radio signals are a small part, where in the spectrum is this radiation to be found? As discussed in Chapter 3, it might operate at extreme ends of the electromagnetic spectrum but the range has been thoroughly searched and no trace of a 'psi-wave' discovered.

If PK is produced by a genuine form of mental energy or acts via some as-yet-unknown force, then the only conclusion to be drawn is that this force has nothing in common with others we have so far experienced in the universe. Depending upon your viewpoint, this fact either strengthens or weakens the case for a paranormal explanation for PK. To the sceptical, it merely reaffirms the claim that because PK cannot be explained by recourse to electromagnetism it must be put down to trickery or sleight of hand. To the believer, it confirms that the phenomenon is not governed by the normal laws of physics, cannot be measured by humans wielding electronic gadgets and lies above and beyond us. Perhaps PK is an adjunct to some odd quantum mechanical effect, but if this is the case, an explanation for the way it works may be a very long time coming.

References

1. John and Anne Spencer, *The Encyclopaedia of the World's Greatest Unsolved Mysteries*, Headline, 1995, p. 259.
2. L.E. Rhine, *Mind Over Matter*, London, Macmillan, 1970.
3. As quoted in ref. 1, p. 261.
4. As quoted in Richard Milton, *Forbidden Science: Exposing the Secrets of Suppressed Research*, Fourth Estate, 1994, p. 46.
5. Hans Eysenck and Carl Sargent, *Explaining the Unexplained*, Weidenfeld & Nicolson, 1982, pp. 102–3.

Chapter 5: The Fire Within

Billy, in one of his nice new sashes,
fell in the fire and was burnt to ashes;
Now, although the room grows chilly,
I haven't the heart to poke poor Billy.

Harry Graham

The earliest reported case of what has become known as Spontaneous Human Combustion (SHC) dates from early in the seventeenth century when a carpenter named John Hitchen died along with his wife and daughter during a thunderstorm. Recovering his body from the wreckage of his house, neighbours were astonished to discover that the man's body was burning from the inside but there was almost no outward sign of fire. He continued to burn for three days until his entire body had been reduced to ash.

Of course reports like this from so long ago are, by their very nature, unreliable and it is interesting to note that until this century, all victims of SHC were recorded as having been heavy drinkers even if they were not. Although the official line taken on SHC by the medical profession has changed little since the first case – that all victims are in some way associated with fires or were alcoholics – there is an impressive body of evidence to suggest there is more to this rare and frightening phenomenon. Coroners do not refer to victims dying from SHC even if all the evidence points that way and correlates with many other similar cases, and perhaps because of its rarity, spontaneous combustion of human beings is still viewed as paranormal and therefore fantasy.

The most famous case of SHC was indeed fantasy – it sprang from the fertile imagination of Charles Dickens who dispatched the character Krook in *Bleak House* (published in 1853) using the device of spontaneous combustion. But in recent times there have been several baffling cases which have been witnessed, sometimes by scores of people, and the intriguing medical details analysed by sceptical experts as well as paranormal investigators. In 1986, the Viennese pastor, Franz Lueger is reported to have burst into flames and exploded before an entire congregation during a particularly impassioned sermon, and in Los Angeles in 1990, Angela Hernandez blew up on an operating table at UCLA.

At first glance, many cases of SHC seem to be easily explained. Victims are frequently associated in some way with naked flames or flammable liquids. There is also a disproportionately high number of victims who are homeless people who, upon further investigation, are found to have been alcoholics. Yet, as with many phenomena categorised as 'paranormal' there are a significant percentage of incidents that completely defy these standard explanations. Furthermore, there are some aspects of SHC that are extremely difficult to accommodate using accepted medical knowledge and may require a reappraisal of some treasured notions of what the human body is capable of when exposed to extreme conditions.

First, let us consider the anomalies associated with SHC.

The most striking aspect of Spontaneous Human Combustion is that in almost all cases the body of the victim has suffered the results of intense fire, but the surroundings are relatively unaffected. In many incidents the body has burned from the trunk outwards, but the fire has not reached the extremities. Some bodies are reduced to ash, but fingers, and on occasion, entire limbs show almost no trace of combustion. Frequently, clothes and shoes are undamaged and items lying close by are unmarked.

The second surprising aspect of SHC is that those parts of the body affected by the fire, including bones of the rib cage, the pelvis and the backbone are often completely reduced to ash. This is significant because the temperature required to powder human bone is relatively high – in excess of 1000 °C. Incinerators used in crematoria operate up to 950 °C and even at these temperatures, bones often have to be crushed mechanically after cremation.

So, explanations for SHC have to account for the very high temperatures required to cause the sort of damage reported and suggest ways in which a human could burst into flames from a source that appears to be internal.

One of the earliest attempts to produce an official explanation for Spontaneous Human Combustion relies on the questionable claim that the vast majority of cases involve an external flame or heat source. It is true that in a large percentage of reported cases there is a local heat source, but certainly not all. And, in a significant number of those incidents where fires or ovens are found nearby, they were not in use at the time of the incident. A famous example is the death of the student Jacqueline Fitzsimon at Halton College in 1985, who burst into flames walking down the stairs leading from her catering class. Her lecturer, Robert Carson, swore that all the rings on the ovens had been turned off an hour before the group left the room and a Home Office chemist, Philip Jones, was unable to make a smouldering catering jacket burst into flames.[1] A subsequent investigation discovered that she refused to wear the regulation jacket and routinely carried a box of matches in her pocket.

Continuing with the argument that heat sources are always present, sceptics explain that combustion is facilitated by victims either falling onto a fire or becoming unconscious and leaning against a heat source. The victim then experiences what has been called the *wick effect* by which the human body acts like a candle, slow burning from the inside outwards.

Fire Safety Engineer Dr Dougal Drysdale described how this worked on a television documentary called, *A Case of Spontaneous Human Combustion*, broadcast by the BBC in 1989.

'In a way, a body is like a candle – inside out,' he said. 'With a candle the wick is on the inside and the fat on the outside. As the wick burns the candle becomes molten and the liquid is drawn onto the wick and burns. With a body, which consists of a large amount of fat, the fat melts and is drawn onto the clothing which acts as a wick, and then continues to burn.' He then went on to demonstrate the effect with a lump of animal fat wrapped in cloth.

It has since been claimed that the demonstration was falsified using time-lapse photography making it appear that the fat had combusted furiously and with little effort. But, such criticisms aside, the demonstration did not begin to address the most significant facts surrounding cases of SHC. Humans may be a little like candles in one respect but in other, rather important ways, they are quite different. Firstly, viewers were not told what sort of fat was used in the experiment. Secondly, no account was made for the fact that the most obese humans are composed largely of water rather than pure fat. But, crucially, the question of temperature was conveniently side-stepped and no serious attempt was made to explain how the wick effect could generate temperatures even close to those required to destroy human bones.

Given then that the wick effect offers little assistance in trying to explain the majority of SHC cases, whether initiated by a local heat source or not, are there any better explanations?

Perhaps a more productive approach would be to look at the type of fire witnessed in many cases of SHC. Although there is a spectrum of effects, one common factor is the presence of a blue flame. Another is the rather surprising fact that water has little effect in quelling the fire and in some cases actually makes it worse.

The appearance of a blue flame suggests the presence of methane gas, (the major component of Natural Gas). Cattle often experience a build up of methane in the digestive tract which bloats the stomach and sometimes requires a vet to make an incision to release the pressure. It has been suggested that a similar process can occur in humans creating pockets of gas which under normal circumstances can be released eventually but in rare cases meet with some intense heat source – with explosive results.

The fact that water facilitates rather than extinguishes the fire is more curious. Fires of this type are usually linked with the presence of the elements magnesium or titanium. The best example is aircraft fires because of the high proportion of magnesium and titanium used in the framework of airliners and military craft. Magnesium burns with a very intense, very hot flame and in the presence of water, it releases hydrogen gas which is itself highly flammable. Magnesium is also one of the trace metals found in the human body.

Yet, these links do not in themselves point the way to how SHC occurs. Magnesium is present in the body, but only as a trace element. It could conceivably act as a 'fuse' to initiate a fire which would then be fuelled by more common materials in the body, such as methane. But so far, no one has provided a demonstrable mechanism via which this could happen.

The chemical reactions that occur in our bodies all the time are varied and incredibly complex. The human body is an elaborate machine which can absorb energy in the form of food, perform a variety of specialized as well as general functions both voluntary and involuntary and then emit the waste after all the usable energy has been extracted. Whether it is the movement of a leg, the mechanism behind the creation of thought or the systems governing breathing, all the functions of the body eventually come down to biochemical processes. Although these processes involve a complex array of biochemical devices, each of them comprises a set of simple, step by step reactions.

It was discovered in the nineteenth century that chemical reactions are either exothermic, that is they evolve heat, or else they are endothermic, they draw in heat from the environment. Biochemical processes are simply chemical reactions involving bio-molecules, large organic molecules found in living things. Some of these reactions are exothermic, others are endothermic. Perhaps, in cases of SHC, these chemical systems become distorted in some way and the energy produced is channelled into precipitating an uncontrolled process.

Whether caused by the ignition of methane or by some strangely distorted exothermic processes, the common objection to any form of fire starting in the human gut has been the presence of large amounts of water, a chemical that constitutes about 70% of body mass. But, recent studies conducted by chemical engineer Hugh Stiles has shown that the presence of water does not actually affect the chances of ignition and the sustainment of such a fire.

The mechanism he describes begins with a heat source oxidizing the carbon of organic molecules found in the cells of the gut to carbon dioxide. This is an exothermic reaction and generates a large quantity of heat which then evaporates the water in neighbouring cells. Calculations show that the trunk of an average human being burning in this way would produce in the region of 200 megajoules of heat (200 million joules), which is more than twice that needed to evaporate the entire water content of the body. It then follows that the left-over energy feeds the process resulting in an intense fire from within or else an explosion sparked perhaps by the ubiquitous flammable gases in the gut.

The problem with this theory is that such a reaction would produce large amounts of water vapour which would then condense on every surface near the site of the incineration. According to documented reports, this is not usually the case. Many cases of

SHC occur when the victim is alone and bodies are often found long after the fire has run its course, by which time any water may have been removed by secondary fires. But in cases where the body has been found soon after the event, or when the conflagration has been witnessed, there have been no reports of unusual condensation near the corpse.

Another chemical which presents itself as a candidate for sustaining an internal fire is phosphine. This is a poisonous gas which combusts spontaneously with air at room temperature. It is sometimes seen as a glowing vapour in marshy areas and given the name 'Will o' the Wisp'. It is not normally found in the human body but could be produced by an unusual reaction involving naturally occurring phosphates.

According to retired industrial chemist Cecil Jones, phosphine could account for some reported cases of SHC and has even gained an official name – the *phosphinic fart*. This may account for some cases but clearly not all.

For a more general source of heat, we need to return to biochemistry, but this time look at the energy-exchange processes that take place in the cell.

Humans gain energy from food by a series of biochemical processes. The first step is to convert food into glucose. This is then subjected to a process called *glycolysis* during which it is converted to a molecule called pyruvic acid. The pyruvic acid is then changed into a variety of molecules and along the way, energy is released. Although this is a complex process overall, each step involves a simple chemical change.

An example is the famous Krebs' cycle named after the biochemist, Hans Krebs which takes place in a part of the cell known as the mitochondria. As the name implies, Krebs' cycle is a 'loop', or a cyclical system, in which pyruvic acid enters a reaction cycle and is converted into a series of other molecules. At each step in the cycle, other molecules enter and leave the system siphoning off energy which is continuously made available by more molecules of pyruvic acid entering the cycle.

The crucial point about this and other biochemical systems is that each step involves the production of a tiny amount of energy, but collectively they provide all the energy required by the body. In order for this source to be used to produce SHC, the energy

Figure 5.1

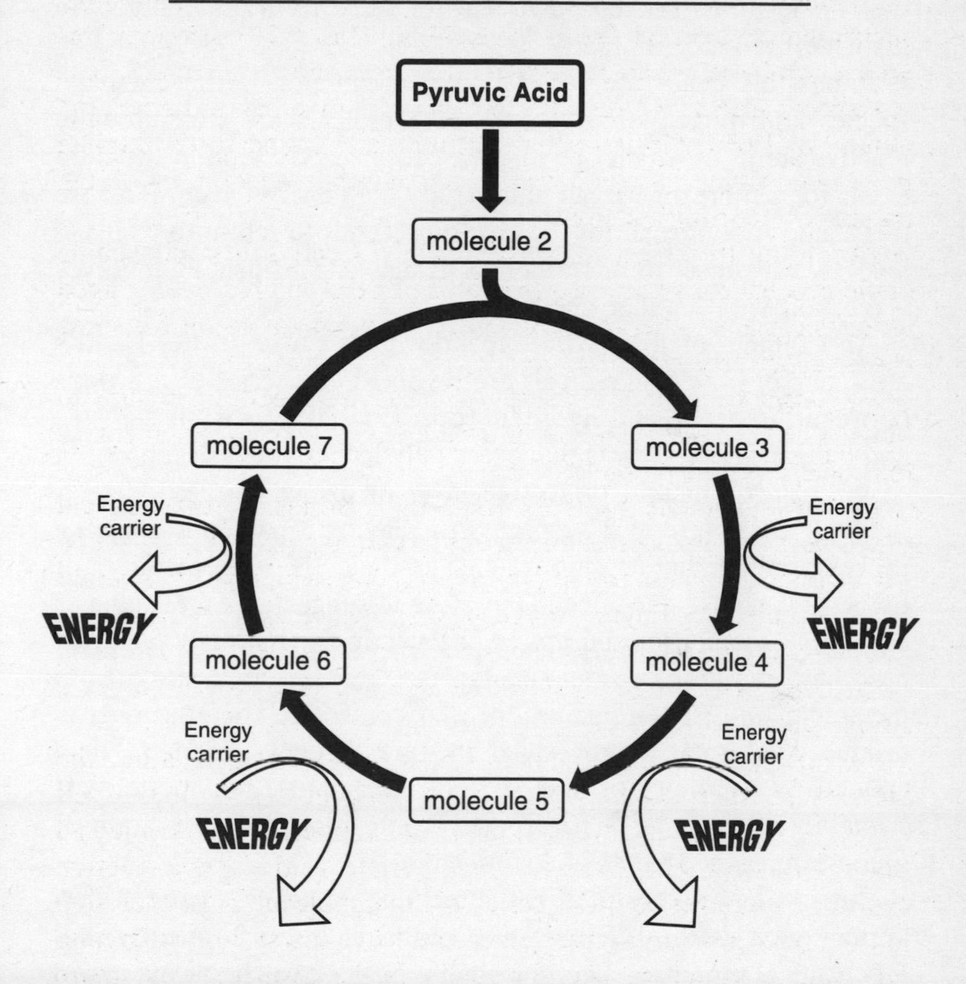
A simplified representation of the Krebs' cycle
Pyruvic Acid
molecule 2
molecule 3
molecule 4
molecule 5
molecule 6
molecule 7
Energy carrier
ENERGY
Energy carrier
ENERGY
Energy carrier
ENERGY
Energy carrier
ENERGY

generated in each step in each cell would need to be utilized in a short space of time.

According to orthodox medicine there is no mechanism by which this has been shown to operate. Each stage of the process is physically separated from the others by tiny membranes within the mitochondria in each cell, but it is theoretically possible for these membranes to be broken down and for some form of 'initiator' to instigate the process.

A possible candidate for the role of initiator is the chemical species known as a free radical. Free radicals are exceptionally reactive atoms in which one of the electrons in the atomic structure exists for a short time in an unnaturally high energy state. Because this state is not normal for the electron, it tries to return to its usual level by whichever means becomes available. This usually involves a reaction with another chemical.

An example of this is the breakdown of ozone in the Earth's atmosphere. Here free radicals are involved in a series of reactions between the ozone and hydrofluorocarbons (HFCs) from aerosols and industrial refrigerants that eventually lead to the decomposition of the planet's natural protective layer of ozone.

Free radicals are also suspected of playing a part in the ageing process, initializing reactions that produce the degeneration of tissue. It is not too far-fetched therefore to suggest that free radicals can operate under special conditions within the body. Whether or not they can initiate the spontaneous release of energy from metabolic processes remains open to conjecture, but the energy produced by such a process, if it were to occur, is more than enough to account for the sort of heat and dramatic effects witnessed in cases of SHC. To put it into perspective, 1 gram of glucose releases 38 000 joules of energy.

But just how likely is it that something could go wrong with a system such as this? The answer is that it does, frequently. All biochemical processes are controlled by elaborate chemicals called enzymes. The complex series of steps that produce the correct type of enzymes in the necessary proportions at the right time are controlled by specific genes. Sometimes these genes malfunction. If this occurs, a wide range of problems may result. For example, if the gene controlling the enzymes responsible for cell-division malfunctions and fails to turn off the enzyme, uncontrolled cell division occurs leading to cancer.

If a similar malfunction occurred in the genes controlling a system such as glycolysis or the Krebs' cycle, it is conceivable that it could get out of control. Instead of a cancerous growth, the body would instead witness greatly increased metabolic processes. Taken to an extreme and conducted over a short time period – perhaps hours instead of months or years, this could provide the mechanism for generating the energies needed to bring about SHC.

An alternative to using the metabolic processes within cells might be to utilize the fat stores which accumulate around the abdomen. This may also account for the fact that in most cases of SHC, the fire is seen to come from the trunk or the midriff.

Hibernating animals, such as squirrels, possess a particular form of fat called *brown fat* which is broken down to supply energy during the hibernation phase. This is an extremely efficient energy source providing in the region of 500 watts of power in the form of heat for each kilogram stored. This is enough to power a small electric fire.

Humans do not possess brown fat, but when the body is starved of nutrients 'normal' fat may be broken down. This is a relatively inefficient process, but an obese individual would have a large stock to draw upon and again, a free radical initiation may possibly enable a form of chain reaction to be sparked off which would readily supply the sort of energy needed to bring about spontaneous combustion.

As I mentioned earlier, the pattern of fire in the victims suggests the combustion begins in the gut and victims are invariably found with their extremities unscathed by the fire. This arrangement would fit with both the mechanisms described – the cellular and the body fat method. The reason depends upon which theory you adhere to.

We should first consider the cellular mechanism. In order for the mitochondria to continue functioning, they need oxygen. After the victim has died, the heart stops beating, but parts of the body live on until the oxygen supply is used up. By the time the fire reaches the knees, the oxygen level in the mitochondria would certainly have dropped below a critical minimum and the cells would die and combustion would stop.

If instead we consider that body-fat is responsible for sustaining combustion, the fire would probably burn itself out by the time it reached beyond the trunk because the extremities of even the most obese person are usually less fatty than the abdominal region.

To round off the list of possible mechanisms, I should mention that in some recent studies, the phenomenon of cold fusion has been mooted as a cause. This strange process was mentioned briefly in Chapter 1 as a possible energy source for interstellar spaceships, but is seen by most of the scientific community as little more than fantasy.

In a set of experiments conducted in 1989, Professors Fleischmann and Ponns devised a system which appeared to produce energy by nuclear fusion at room temperature. The process seems to share some of the characteristics of what are called electrochemical systems. In these, reactions on metal surfaces immersed in special solutions provide electrical energy: this is the basis for the common battery. Rather than employing the more usual chemical reactions that take place in electrochemical devices, the two scientists proposed that, under the right circumstances, metal surfaces could act as facilitators for the fusion of atomic nuclei.*

Professor Fleischmann maintains to this day that his system works, and his attempts to create a workable and practical energy source from cold fusion have been funded by Toyota for the past five years. How this mechanism could create SHC is unclear, but in their book *Spontaneous Human Combustion*, authors Jenny Randles and Peter Hough quote an anonymous physicist as saying:

> It seems likely that of the elements found in the body, potassium or sodium are the most likely to be involved in a fusion reaction. The tissues with the highest potassium levels are in the brain, the spinal cord and skeletal muscles. This would suggest that the feet and the hands would be the least affected, and the head the most.
>
> It would be very hot, maybe incandescent. This would hardly affect surrounding materials, as a slow burn seemingly would. It would all be over in seconds, or a few minutes at most. There just might be residual

* Chemical reactions involve the electrons of an atom rather than the nuclei. Atomic processes, both fission and fusion, employ the nuclei and have nothing to do with these extranuclear electrons. Because far less energy is needed to keep electrons in atoms compared to that required to keep neutrons and protons in nuclei, chemical reactions are much easier to facilitate.

radioactivity, it would be surprising were there not, but not necessarily a lot.'[2]

This is an interesting proposition and sounds convincing, but the problem is that if we accept this as a cause, we have actually travelled further away from an explanation. Not only is SHC unproven, cold fusion is regarded by the majority of scientists as nothing more than a hypothesis which has been shown to be unrepeatable by almost every worker in the field.

There is really no need to rely upon explanations from eccentric regions of science to account for ways in which SHC could conceivably happen – there are plenty of possible anomalies within the biochemistry of the body and the natural forces around us to generate just-plausible mechanisms.

One SHC researcher, Larry Arnold, has offered up the 'strange particle' hypothesis as a source, suggesting the involvement of an imaginary particle he calls the *pyrotron* which could in some mysterious fashion initiate combustion. Although, like the *psitron*, the pyrotron has a cute name, there is no experimental or theoretical support for its existence and it is, I believe, a quite unnecessary complication.

So far I have concentrated on ways in which the body could sustain a form of combustion once it has been initiated, but what sort of event could spark off the process?

The first place to look should be the body itself. Although this sounds an unlikely source for the sort of energies needed, it is not beyond the bounds of reason. We saw in Chapter 3 that the human brain is a mass of electrical circuitry more complex than any computer so far built. Each neuronic impulse operates at a voltage of 100–120 millivolts. It is possible that a suitable current, perhaps produced by a surge of energy from some unknown freak activity somewhere in the brain, if re-routed, could initiate combustion in a minute area which could then lead to a series of increasingly disruptive processes ending in an explosion or an internal 'fire'.

The biochemical processes taking place in cells also involve electrical potentials. A voltage higher than that found in the brain is produced across the inner membrane of the mitochondria – a potential difference of 225 millivolts. According to one calculation, this amounts to 45 000 volts per cubic centimetre.[3]

But we should not get carried away with such figures. Adding together voltages from millions of neighbouring cells does not actually mean very much. It is a little like saying that each computer in a building handles a voltage of 240 volts and if this malfunctioned we could create an electrical surge of tens of thousands of volts by it spreading to all the hundreds of other computers nearby. It is conceivable that such a peculiar circumstance might arise, and in the same way it is possible that a freak event could occur inside a human body, but it would be a very unusual anomaly.

Looking beyond the body itself, there are powerful natural and manmade forces at work. It has been estimated that 16 million thunderstorms occur on the planet each year and at any one time there are 1900 of them raging in different parts of the world. Surprisingly, only a handful of people are killed each year by lightning in Britain (although in America with a population only four times larger, 100–150 individuals are killed by lightning annually).

Lightning is produced in thunder clouds, which are churning masses of air containing large numbers of negatively-charged ions. These ions repel negative charges in objects over which the cloud passes, disturbing electrical fields in all living creatures, including humans. This is why animals and some sensitive people know when a thunderstorm is approaching. Some individuals have been known to feel depressed or even physically sick as the negatively-charged cloud moves overhead. The brain chemistry and the electronic balance in the cells of the body are disrupted by the repulsion of negative charges, and in some unknown way, this causes emotional and physical responses.

According to reports, there is no apparent correlation between thunderstorms and Spontaneous Human Combustion. When people are struck by lightning they do not burst into flame. In fact, most victims show no outward sign of having been burned. The reason for this is that the intense heat produced by a thunderbolt, (something like 30 000 °C) touches the body for only a few microseconds which is not long enough to cause a burn. Instead, the victim dies from a complete destruction of the nervous system – the electronic circuitry of their bodies is literally fused.

Although there is no direct correlation between lightning storms

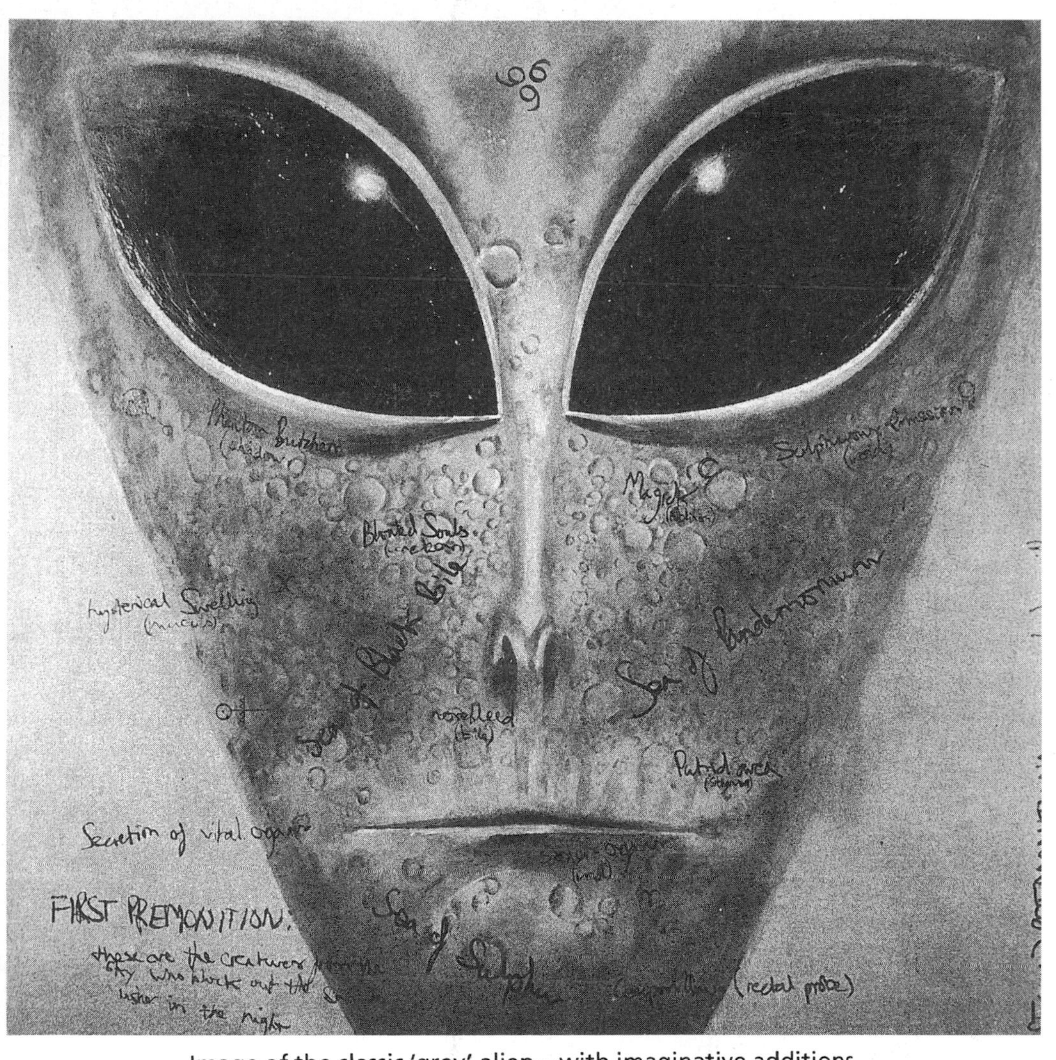

Image of the classic 'grey' alien – with imaginative additions
(Rod Dickinson/Fortean Picture Library)

Baron von Schrenck-Notzing conducting a psychokinesis experiment in the 1920s *(Fortean Picture Library)*

A photograph of the legendary Bigfoot, taken in California in 1967 *(René Dahinden/Fortean Picture)*

A 16th Century map, warning sailors of the locations of sea monsters
(Fortean Picture Library)

The remains of Dr. John Irving Bentley, who died of spontaneous combustion in Pennsylvania, USA, in 1966 *(Copyright 1976/1993 by Larry E. Arnold (Fortean Picture Library))*

A Coventry Freeman's Guild dinner in 1985. Look in the top left-hand corner for the uninvited, ghostly guest *(Haddon Davies/Fortean Picture Library)*

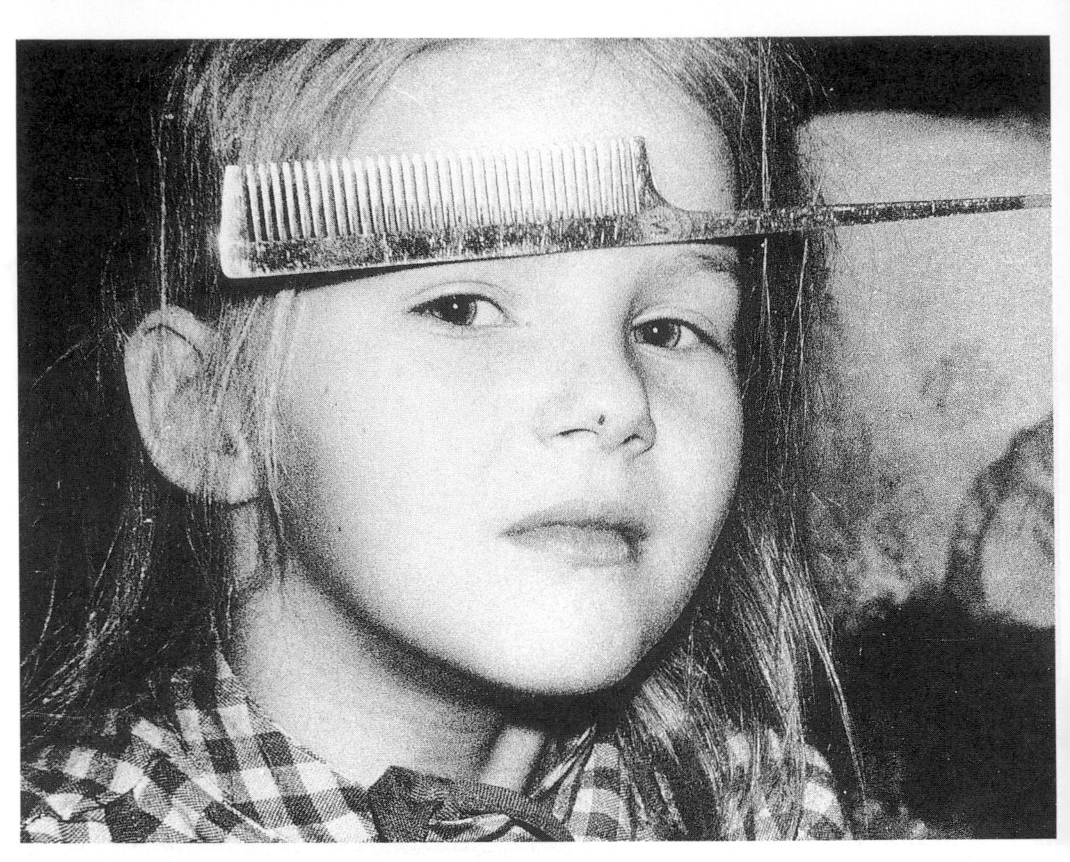

Above: Visionary, poet and artist William Blake's interpretation of the near-death experience (painted as an illustration to the poem 'The Grave' in 1812)
(Charles Walker Collection/Images)

Opposite above: Svetlana Glyko, an 8 year-old Russian girl who attracts small metal objects. Photographed in the early 1990s *(Fortean Picture Library)*

Opposite below: 'And I beheld when he had opened the sixth seal, and lo, there was a great earthquake; and the sun became black as sackcloth of hair, and the moon became as blood'

REVELATIONS 6, 122

'The opening of the Sixth Seal' by Francis Danby, engraved by Nicholls.
(Charles Walker Collection/Images)

Fire walkers on red-hot embers during the Chinese festival of the Great Monkey God in Hong Kong *(Images Colour Library)*

and SHC cases, there are similar natural weather effects that could account for some reports. *St Elmo's Fire*, a ghostly glow sometimes seen around the masts of ships and tall buildings, is produced by ionized air.

A similar phenomenon is ball lightning. Used frequently to account for a collection of paranormal activity, from UFOs to crop circles, it is a little understood weather effect and so rare its very existence is still disputed by many meteorologists. If it is a genuine weather feature, the consensus from occasional sightings is that it takes the form of a ball of energy which is generally spherical, from 1cm to more than 100cm in diameter that lasts for a few seconds. The balls are reported to move horizontally at speeds of a few metres per second and to decay silently or with a small explosion. It may be a form of condensed energy, perhaps a small cloud of ionized gases that sometimes breaks away from a thunder cloud, imagined by some to be 'thunder-bolt-in-a-ball'.

Ball lightning does present itself as a serious candidate for initiating SHC. It is very rare, it possesses large amounts of energy, appears, from some accounts, to be able to penetrate material objects and lasts for a very short time. Furthermore, it leaves no clear trace of its presence.

Another possible initiator is the increasing number of electronic fields produced by the machines we use. It is easy to forget that every piece of electronic hardware in our homes, offices and cars have associated magnetic fields and each, to some degree, interfere with one another.

I remember as a child how the television picture in our living room would become disrupted when the neighbour turned on the lawn mower or the food mixer, (it was always in the middle of *Star Trek*, I seem to recall). Today electronic machinery is better insulated and this cross interference is rare, but even the best insulation cannot prevent a certain amount of leakage and as we fill our environment with ever more sophisticated machines, the chances of stray fields and the possibility of power surges altering the electrochemical make-up of our bodies should be considered seriously. Much has been made of the eye-strain resulting from the use of computers, yet little research has been conducted on the possible detrimental effects they have upon our nervous systems.

Static electricity is also with us on a daily basis and some people are far more susceptible to it than others. There have been many cases of people who act as stores for static electricity and have even unconsciously disrupted electronic devices nearby. It is conceivable that victims of SHC are particularly vulnerable to electrical fluctuations in the environment generated not by natural weather conditions such as thunderstorms but by the machinery around them.

Beyond these relatively mundane sources, there have been a collection of what might be called 'paranormal explanations' to account for the initial 'spark' that triggers SHC.

Many of these musings have no theoretical grounding and appear to be unnecessarily mysterious, yet one interesting fact has arisen from them. It would appear that spontaneous combustion only occurs in human beings.

It may be that SHC is such a rare phenomenon that we have simply not observed it in the wild, but reports would be expected from farmers through the centuries, especially given the propensity for cows to store large volumes of combustible gas. The cellular mechanisms described above are almost identical in all mammals, so why should they only malfunction in humans?

One suggestion is that the mood or the psychic state of the victim can somehow influence the chances of SHC. There is no solid evidence to support this, only a few incidental facts that could point to some form of emotional trigger. Two reports from the 1950s tell of apparent SHC victims who were in the process of committing suicide at the time of the combustion.[4] Some researchers have also suggested that the large number of SHC cases in which the victim was found to have been drunk might be linked to mood rather than biochemistry.

From the collected reports taken over three centuries, there emerges no real pattern to pin down a type of person or any particular peripheral event which may be the root cause of the catastrophe. There is no particular time of day when these conflagrations occur more or less frequently, although more cases are reported during the winter months than the summer. Sceptics would say that this is the time of year when there are more likely to be open fires roaring in the grate or ovens left on. Believers counter with suggestions of a link between SHC and SAD – Seasonal

Affective Disorder. But again, there is no hard evidence to support this connection.

So, what conclusions may be drawn from the supposition, theorizing and meagre collection of facts surrounding SHC?

There seems to be little need to search for any supernatural cause; the human body is a startlingly complex and formidably versatile machine which still turns up surprises for biologist, chemists and physicists. It is possible that an environmental factor could trigger a completely unknown biochemical disturbance causing the body to turn against itself or to become overloaded in some odd fashion.

It is conceivable there are many types of spontaneous combustion. The incidents involving explosion may be fundamentally different to the slow-burn cases or the cause may be the same, but different victims may respond to the initiator in different ways, perhaps dependent upon the quantity of gases in their gut or their physical characteristics. There certainly seems to be a category of SHC which may not be that at all, but are due simply to the build up and eventual release of gases such as phosphine which then combust upon contact with air.

Because SHC is so rare, it is difficult to construct a definitive explanation. The glib responses of traditionalists are certainly unsatisfactory and account for only a small number of cases, ignoring aspects of many others. In this respect the 'scientific' explanations are not scientific at all. They merely sweep under the carpet any uncomfortable facts – those that defy explanation by conventional means. Sadly, these 'difficult' aspects are also the ones which offer the best opportunity to discover something strikingly new.

References

1. John Fairley and Simon Welfare, *Arthur C. Clarke's Chronicles of the Strange and Mysterious*, Collins, 1987, p. 160.
2. Jenny Randles and Peter Hough, *Spontaneous Human Combustion*, Bantam, 1993, p. 238.
3. John. E. Heymer, *The Entrancing Flame*, Little, Brown, 1996, pp. 170–1.

4. Michael Harrison, *Fire from Heaven*, Scoob Books Publishing, 1990.

Chapter 6: Visions From A Future Time

For we know in part, and we prophesy in part.

1 Corinthians 13.9

Of all paranormal phenomena, probably the most difficult for the scientist to accept is precognition. Ironically, it is also one of the most ancient metaphysical ideas and appears in some form in almost all cultures and eras, adopting many guises from Shakespeare – 'Beware the ides of March,' to end-of-the-pier novelty attractions.

Before we look at the ways in which precognition may operate via some rather elaborate ideas from modern physics, let us first consider the more mundane ways in which some forms of foretelling the future can be explained.

In his book *Supernature*, Lyall Watson lists over a dozen exotic forms of prediction, ranging from aeromancy (divination via cloud shapes) to tiromancy (a form of prophecy involving cheese), but there are more traditional forms of fortune-telling and wide-scale prophecy.[1]

In ancient Egypt the priests of the sun-god Ra were prophets and the Greeks consulted oracles such as the one at Delphi to determine the most auspicious time to wage war or to broker peace. In more recent times, the most famous western prophet was Michel Nostradamus, and methods such as the tarot, crystal ball-gazing and palmistry have become part of cultural history, embodied in the literature and folklore of many countries and eras.

Nostradamus was a mystic who lived in France between 1503 and 1566 and composed thousands of prophecies in the form of four-line verses called *quatrains*. Some enthusiasts see accurate prophecy within his writing, but Nostradamus is open to the same criticism as almost all traditional forms of divination – the problem of interpretation. What Nostradamus predicted has only been fitted with events *after the fact* which rather diminishes their impact, and, like many other forms of prophecy, his statements are extremely vague. Although he has been credited with predicting many technological discoveries, wars, dictators and catastrophes of the twentieth century, the original prophecies hardly offer clear-cut pronouncements. Take the following quatrain for example:

> They will think they have seen the Sun at night
> When they will see the pig half-man:
> Noise, song, battle, fighting in the sky perceived,
> And one will hear brute beasts talking.
>
> (Century I, verse 64)

This is meant to be a prediction of atomic weapons and fighter aircraft, but it is easy to superimpose such images after the events have transpired. Commentators of the nineteenth century may have placed a totally different interpretation upon it and if the work of Nostradamus had not been discovered until say the twenty-second century, what would people living in that century think of it? Like many occult phenomena, prophecy or precognition is judged by the standards and universal viewpoint of the time.

A quite different form of precognition is that involving an individual who is interacting with a system, for example, practitioners of the Chinese divination method called the I Ching or the European version, the tarot. Nostradamus was a lone prophet, sitting in his study supposedly gazing into the future, but tarot or the I Ching are a form of ritual which, believers claim, focuses certain mental powers and bypasses the conscious mind. Enthusiasts of these techniques suggest it is our subconscious minds that can break the shackles of time and that, if we allow it free rein, we have the power to access the future. By using any system, whether it be a set of cards or the yarrow stalks of the I Ching, the everyday clutter of the conscious mind is filtered out so that the subconscious can take direct control.

In a similar way, many individuals who claim to have the power of precognition say they receive images from the future while they sleep. If there is any truth in the phenomenon, then it would make sense that the best time to receive messages would be during sleep as this is when the subconscious takes over from the conscious mind.

Whether prophetic images come via a ritualistic system or by drug or sleep-induced states in which the conscious mind is bypassed, a surprising number of people claim to have had precognitive experiences. Most of us forget the images long before we might try to interpret them – they are crowded out by the conscious mind and the confusion of everyday life. But, according to believers, some of these images remain with us and pop up again in our conscious minds when triggered by some subliminal event – this is what is usually meant by a feeling of *déjà vu*.

Some people who retain images from dreams can consolidate them and a few have even been so bold as to make public their predictions. Surprisingly, these have, on occasion, proven disturbingly accurate. Worst still for the scientist, once in a while, these precognitive images are almost impossible to dismiss with mundane, mechanistic explanations.

After the Aberfan disaster in 1966 when a coal slide destroyed a school, killing over one hundred people, many claimed they had 'seen' the event before it had happened. According to one psychiatrist who studied the cases of precognition associated with this disaster, some sixty predictions appeared to him to be genuine. One of the children who perished in the tragedy told her mother two weeks before the disaster:

'No, Mummy, you must listen. I dreamt I went to school and there was no school there! Something black came down all over it!'[2]

One William Klein foretold the sinking of the *Titanic* and was so impressed with his own vision of events to come that he warned a friend not to travel on the ship. The friend ignored his warning and met an icy death. A well-known researcher of the paranormal at the turn of the century, W. T. Stead was warned by two psychics not to travel on the *Titanic*, but surprisingly, he ignored both of them and drowned.

These claims could of course be put down to coincidence. It is not unusual for children to have anxiety dreams linked to school and this could take the form of something black engulfing the building. The *Titanic* was a ship of revolutionary design for the time and there was a great deal of publicity surrounding the maiden voyage of what was claimed to be the first 'unsinkable' ship. It is not surprising that many would have thought the worst. When the worst happened it seems as though they had performed a miracle by foreseeing it.

However, some researchers and believers in precognition claim there is no such thing as 'coincidence'. Instead, they claim that human beings have an innate ability to roam the highways of time, and that the barriers of 'past' and 'future' may be elastic.

Carl Jung coined the term *synchronicity* and wrote a book about the idea called *Synchronicity: An Acausal Connecting Principle.*[3] In a further collaboration with the physicist Wolfgang Pauli, they created what became known as the Pauli–Jung theory in which they described the unprovable, but nevertheless attractive idea that the collective unconscious could somehow influence the world. They visualized a system which opposes the randomness of Nature by imposing a *unitary world* or an *unus mundus*, guided subconsciously by a universal intelligence (that of the human race and other sentient beings). The archetypes mentioned in Chapter 3 are a manifestation of this enveloping 'force' and a way in which it leaks into the material world via our subconscious minds.

This is an appealing idea, rooted in humanism and transcending the mundane world in which we live and work, but is there any substance to it? How could such an all-pervading awareness operate within the rules of science?

If we restrict ourselves to the matter of predicting the future, there are two possible mechanisms via which this might happen. Both require the use of ideas from the very fringes of modern physics, but which are not illogical or, in theory, beyond the limits of recognizable science. The first of these depends on that useful theoretical construct – the wormhole.

We saw in Chapter 1 how wormholes might be used to travel large distances across the galaxy or beyond without having to traverse the distance between two far-flung points. There are huge practical problems involved with this method of transport and I showed how it could only offer, at the very least, a limited form of communication between two fixed points. But, as well as offering a passage between two physical points within the same space-time framework, wormholes may also offer an opportunity to access information from the future.

The wormhole in Fig. 1.4 linked two points in space, but if you now picture one end of the wormhole moving at close to the speed

Figure 6.1

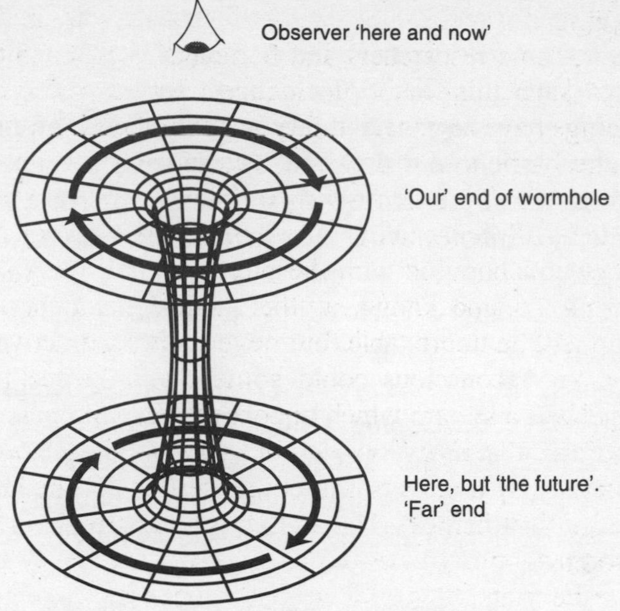

of light (Fig. 6.1), the whole system suddenly takes on the characteristics of a time machine.

One of the consequences of Einstein's special theory of relativity is that the measurement of time is dependent upon the velocity of an observer. If one end of the wormhole is moving at close to the speed of light whilst the other is moving much more slowly, then observers at the two ends would be in different times.

For our precognition scenario, we need to have the far mouth of the wormhole positioned in space not far from where we are at the other opening, so as to 'see' events that have meaning for us, here on Earth. But, our end of the wormhole would initially have to be moving at a velocity close to the speed of light for there to be any viable difference in the measurement of time at the two ends. But once the distinction between the ends is established, the velocity difference can be zero.

If the other, far mouth of the wormhole was relatively stationary compared to ours, it might follow a circular path once in 7 days but to us it would seem to take far less time, let us say, one day. So effectively, we have a time machine – the far end lies in our future.

An observer at 'our end' looking through the wormhole to another observer at the far end would see that their date was six days in our future and we would exist in their 'past'.

So how could this facilitate precognition?

Prophecy or precognition requires the passage of information from a future time to our time. To do this there is no need for the complexities of transporting a spaceship or people through treacherous wormholes with their associated gravitational fields and devastating forces. If information can pass through the wormhole and can be detected by a sensitive individual, then precognition is a possibility.

In his book, *The Physics of Star Trek*, Lawrence Krauss analyses the possibility of developing a matter transporter as used in the television series *Star Trek*.[4] His conclusion is that the transportation of a material object is practically impossible, but that the system might just work if we view the transporter as moving *information* rather than bodies.

Even then, the mathematics is enough to induce a migraine! It appears that the information content of a human being – all the data needed to reproduce a single human – is in the region of ten million billion times the total information contained in all the books ever written. But a way around this restriction is to say that information conveyed from a future event through a wormhole to a receiver in our time would not need to be so detailed.

Documented cases of precognition demonstrate one unifying characteristic – the images from the future are almost always nebulous and ill-defined; often subjects receive a mere outline describing the event. This then may require a far less sophisticated information transfer system. Psychics who 'see an outline of a ship sticking out of the water' and later claim to have witnessed the sinking of the *Titanic*, may have received information from the event which somehow leaked through a wormhole linking the future with the present; they then processed that information and 'saw' the catastrophe before it occurred.

When outlining the use of wormholes for space travel, I highlighted the fact that they would be of limited use. This is also the case when using them to explain precognition. The far mouth of the wormhole would have to be in just the right place at the right time in order for information about an event to be passed back to the receiver. This in itself seems highly improbable.

Believers in this theory have suggested that perhaps mini-wormholes are present within our environment all the time and that in some inexplicable way, catastrophic events trigger them to act as conduits for information. They go on to suggest that human consciousness can somehow interact with these mysterious, theoretical objects and when the human mind faces catastrophe, or perhaps death itself, information can be passed through a wormhole linked to our present. But, until we know more, this explanation is actually no explanation at all, merely an hypothesis used in an effort to add substance to another. Instead many believers in precognition are beginning to show an interest in the idea that prophecy may be linked with that much-maligned area of physics known as quantum mechanics (QM).

Quantum mechanics began life during the early part of the century and came as a completely revolutionary theory that overturned the prevailing (classical) ideas of Victorian physicists.

The earliest model for the atomic world held that the atom was composed of a nucleus around which electrons orbited – like a solar system in miniature. Electrons were known to have about one two-thousandth the mass of a proton (one of the constituents of the nucleus) and to possess a negative charge to counter the positive charge of the proton.

But during the first decades of the century it was realized this model could not possibly work. For a start, mathematics demonstrated that electrons could not be sustained in their orbits like planets and their orbits would decay so that they merged with the protons in the nucleus. As this was clearly not happening in the universe we live in, it was assumed correctly that the model must be wrong.

Through the pioneering work of physicists such as Planck, Bohr and Schrödinger, a far more sophisticated model of the nature of the sub-atomic realm emerged and with it a number of counter-intuitive consequences that have caused confusion for the non-physicist ever since. One of the pioneers of quantum mechanics, Niels Bohr, even went so far as to say that: 'Anyone who is not shocked by quantum theory has not understood it.'

The problems really began when particle physicists realized the electron was not a ball of matter with a negative charge but could only be described in terms of probability. In other words, there is a high probability that an electron will be found a certain distance from the nucleus and a low probability of it existing much further or much nearer the nucleus than this.

Linked with this notion is the Uncertainty Principle developed by Werner Heisenberg in 1927. This shows that there are limits placed upon the accuracy to which pairs of physical quantities can be measured. For example, if we try to measure say the position *and* the momentum of a subatomic particle the very act of measuring these quantities disturbs the particle so much that they cannot be said to have an exact position and an exact momentum at the same time. We can only assign *fuzzy* values for the two factors. This fuzziness is described by the *wave function* – meaning a description based entirely upon probabilities.

Now at first glance this might seem like a trivial matter – so what if subatomic particles cannot be pinpointed to an exact point? In fact this is the whole essence of QM and lies at the root of all the problems it presents for the non-physicist. It is also the very reason why quantum mechanics could conceivably help to explain precognition.

If we cannot define the universe at the most fundamental level, then it must mean the universe is constructed upon probabilities. There can be no certainty, no clear-cut definitions, no pure 'yeses' or 'nos'. From this spring some very strange quantum mechanical ideas.

The first of these is that the universe can only be studied on a statistical level; if we probe too deeply, or try to single out individual particle transactions, we end up with nonsensical results. An alternative way to consider this is to say that the universe really only demonstrates an apparently logical framework if viewed holistically.

This we can just about accept intuitively, but modern quantum mechanics has gone much further. Heisenberg himself suggested that the fuzziness of the quantum world could mean the traditional notion that *effect* must always follow *cause* could break down. Worse still, the experimenter or observer may be able to interfere with the experiment, i.e. that human consciousness might control the processes of the universe in some way.

To illustrate this, consider a famous thought experiment devised by one of the founders of quantum mechanics, the German physicist Erwin Schrödinger during the 1920s.

Schrödinger imagined a box containing a cat and a radioactive source. If this material decays, it triggers a poison that kills the cat, but because the radioactive material decays randomly, there is a fifty-fifty chance it will happen and kill the cat and a fifty-fifty chance it won't. The only way the experimenter will know what has happened is to open the box to see if the cat is alive or dead. This then means that until then the cat is both dead and alive. The probability will only become a certainty by the action of the experimenter opening the box, so the observer controls the outcome, or in technical terms, 'collapses the wave function'.

Although there is nothing illogical or mathematically false about this description, it does not feel 'right'. Indeed, it leads to a whole series of paradoxical arguments. For example: What if we replace the cat with a human? Presumably they would be as able to collapse the wave function as readily as the experimenter. What would they experience inside the box? Would they be able to override the effect of the experimenter?

Now imagine the experiment had become a media event. What would happen after the experimenter had opened the box? He may find either a live human or a corpse, but the cameras and journalists outside the lab, unaware of the events inside, would not know what has happened, so is the subject alive or dead? Equally baffling is the question of what would happen if the cat or the human was replaced by a computer, or indeed if the experimenter was replaced by a computer? How would these changes affect the outcome?

As mystifying as Schrödinger's cat experiment may be, it is based upon sound theory and decades of reasoning within the discipline of quantum mechanics. It does not feel comfortable because it appears to contravene the logical processes we have been educated to appreciate and some that may be instinctive to us as humans. Yet these principles may be right and our intuition wrong.

These bizarre notions have been interpreted in a number of different ways. The most traditional approach is called the Copenhagen Interpretation which was developed in 1927 and suggests that the indeterminacy of individual events within the subatomic world cannot be extended to the macro world and that the large scale world of everyday experience is only comprehensible on a broad statistical level. Critics claim this side-steps the issue and have come up with some rather lateral ways of looking at the quantum dilemma.

One suggestion known as the Everett-Wheeler model (after the creators of the idea, Hugh Everett and John Wheeler) postulates that all possible outcomes of a process (in the case of Schrödinger's cat – two: dead cat, alive cat) are observed somewhere. In our universe, only one outcome is observed, perhaps the cat survives, but in a parallel universe, the opposite is observed – the cat is dead.

This would mean that whenever anything happens, there are at least two alternatives which are observed by observers in two totally separate universes that can never meet. Furthermore, with each passing second, from the dawn of time, the number of possible 'futures' has increased and is still increasing to a near-infinite variety of outcomes.

This idea is taken seriously by some members of the scientific community. It has been the subject of several papers and used as a platform for further, even more convoluted schemes. Yet it is quite untestable and will probably never be validated. Ironically, its co-creator, the eminent physicist John Wheeler, has attacked parapsychologists, declaring that the scientific community should throw out the 'weirdos from the workshop of science'.[5]

Enthusiasts of the paranormal see this as arch-hypocrisy and claim that the ideas of Wheeler and many other physicists are far more extreme than the most imaginative notions of the parapsychologists and researchers of the occult. The quantum physicists counter this with the argument that their theories are founded on mathematics and a self-consistent body of knowledge stretching back the best part of a century. They also point to the many aspects of quantum mechanics that have been shown to work and the technological developments this discipline has offered the world.

And they have a point. Quantum mechanics is the foundation of the science of lasers, advanced electronics and telecommunications. Without an understanding of this exotic area of physics there could be no television, advanced computing, space travel, CD players, no global telecommunications, no Internet, no laser surgery. By comparison, parapsychology has continued to be one of the most evasive aspects of human exploration and remains impossible to pin down.

A third interpretation which seems less controversial than the Everett-Wheeler model but is nonetheless still beyond scientific verification is the idea that human consciousness can interact directly with the wave function. This suggestion is credited to the American physicist Eugene Wigner and offers the idea that the human mind can subconsciously manipulate the universe at a fundamental level.

Here we have echoes of several ideas discussed in earlier chapters. The parallels with Carl Jung's *unitary world* or *unus mundus* and Rupert Sheldrake's Morphic Resonance are obvious. Such an explanation may account for PK, and types of telepathy, but above all, the Wigner interpretation offers a solution to how precognition might occur.

Central to this explanation is the idea that the interaction between consciousness and fundamental processes is not restricted to the here and now. In other words, human consciousness has the ability to transcend distance and time in the way it manipulates the universe. In technical jargon, it is said to be *spatially* and *temporally invariant*. But what is surprising is that this is not merely the eccentric imaginings of over-enthusiastic believers. It may be supported by a convincing set of tests based upon a thought experiment devised by the American physicist, John Bell.

In the experiment, two particles from a common source are fired at a device which then sends them in opposite directions at the speed of light.

One of these particles is then 'altered'. What this really means is that the particle can be changed in a very limited number of ways. For example, its *spin* can be altered, or if it is a high-energy electron, it can be allowed to return to its normal 'ground state' energy level. But the truly remarkable consequence of this is that the second particle, which, remember, has been travelling in the opposite direction at the speed of light, is also altered *simultaneously* by the change made to the first particle.

For the quantum physicist this presents an exciting but worrying paradox. The speed of light is a finite value, an irrefutable universal constant that cannot be negated; so how could one particle alter the state of the other when any communication between them is seemingly impossible?

Finding an answer to this has puzzled physicists ever since the test was first devised during the 1960s. There are a variety of explanations, but the most popular seems to be the idea that if any sub-atomic particles have once been together, they retain a permanent 'affinity' which appears somehow to transcend physical limitations. In his best-selling book, *In Search of Schrödinger's Cat*, John Gribbin says of this paradox:

> They [experiments based upon the Bell test] tell us that particles that were once together in an interaction remain in some sense parts of a single system, which responds together to further interactions. Virtually everything we see and touch and feel is made up of collections of particles that have been involved in interactions with other particles right back through time to the Big Bang.[6]

And so in this do we have an explanation for precognition and perhaps a form of race telepathy?

The Bell test has actually been demonstrated to work in a range of real experiments conducted since it was first contrived over thirty years ago. But, the problem with linking it to the paranormal is that no information is transferred in the connection between the

Figure 6.2 Bell's test

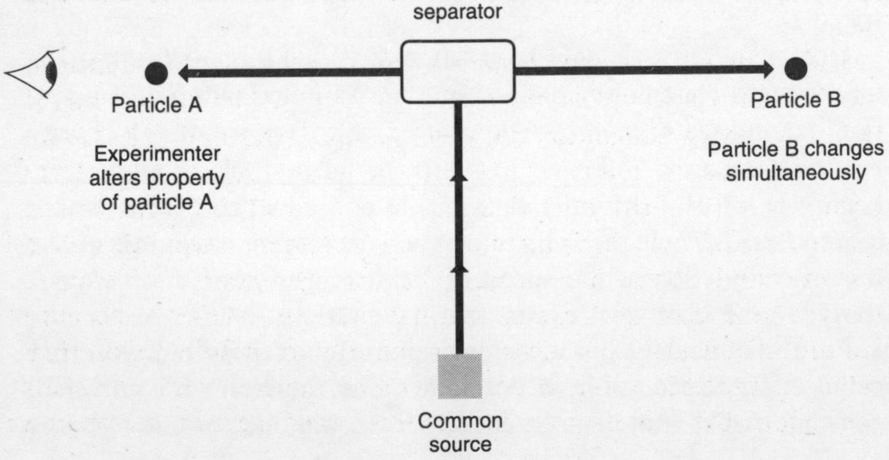

separator

Particle A

Experimenter alters property of particle A

Particle B

Particle B changes simultaneously

Common source

two particles, A and B. If particle A has a certain spin and is 'flipped' into a different state by the experimenter, then particle B is altered in the same way, but crucially, no information need be transferred in order to do this. This change is merely a shift from one spin to another and each spin state is dependent upon a set of what are called *random quantum fluctuations*. Changing from one random pattern to another requires no information. Furthermore, it is found that if an experimenter tries to repeat the test by changing a parameter that does require information to be transferred, there is no simultaneous effect upon particle B.

Theorists of the paranormal say this is actually no barrier because psychic effects do not involve information as we know it, but they cannot support this claim with hard facts and such comments are just the sort that justifiably irritate physicists like John Wheeler.

So what may be judged from this collection of ideas and theories? Several eminent scientists have been convinced that, within certain limits, quantum mechanics can explain the esoteric world of the paranormal. Men such as Jung and Pauli were fascinated with a possible link between the two and constructed elaborate explanations for a marriage of QM and psi. Today, there are many scientists around the world who are continuing the tradition. Henry Stapp at the University of California, Berkeley is utilizing the ideas of Eugene Wigner to reach a mathematical interpretation of how QM could lie behind psychic phenomena and Nobel Laureate Brian Josephson has said that if psi phenomena had never been noticed, quantum mechanics would have predicted them.

Unimpressed by apparent restrictions such as the limit to information transfer, parapsychologists suggest a mechanism for how QM and precognition could relate. They propose that human consciousness is able to influence the past, present and future because, at its most fundamental level, the law of causation can be manipulated. They claim the universe is a single unit and that each human mind is part of this vast network. Human consciousness, they believe, can 'see' events that have not yet happened because we are all one, always have been and always will be. This may seem supportable using experiments like the Bell test, say their opponents, but it flies in the face of the Copenhagen Interpretation and ignores the maths while at the same time conveniently extracting the kernel of verifiable science. Furthermore, no one has even begun to offer an explanation for how this vague but intriguing concept could actually work.

In some respects, it is easy to sympathize with the sceptics who say that such interpretations come from ignorant non-scientists who are trying to manipulate something they don't understand to their own ends. To many physicists, the ideas of the parapsychologists are insulting and an affront to the decades of study and research they have committed to their subject. But, on the other hand, quantum mechanics does offer itself as a sacrificial lamb.

To be accepted, even the most unorthodox scientific hypotheses have to be supported by rigorous mathematics and must be consistent with a body of scientific knowledge stretching back to the seventeenth century. Yet, many aspects of modern physics remain unprovable. For instance, there has never been absolute proof of the existence of a black hole and many advanced theories from particle physics remain untestable because the equipment needed to demonstrate a range of effects is beyond our technological capacity.*

To the enthusiast of the paranormal, science at the limit can sometimes seem as esoteric and other-worldly as their own ideas. The difference lies in approach: science is based upon mathematical integrity and where possible, experimental verification – no theory is accepted until proven by experiment. Parapsychologists, the physicists claim, are too easily lead into assumption and neat ideas that appear to fit the facts without proof, mathematical rigour or a sufficient degree of self-criticism.

One of the greatest scientists who ever lived abhorred quantum mechanics despite the fact that he helped to create it. Albert Einstein saw the avenue along which QM progressed as a voyage into the absurd, and commented famously that 'God does not play dice', meaning that the universe was not merely a collection of probabilities. On another occasion he commented: 'Quantum theory reminds me a little of the system of delusions of an exceedingly intelligent paranoiac concocted of incoherent elements of thought.'

Erwin Schrödinger was as dismissive of modern QM as Einstein,

* Although there is very, very good circumstantial evidence for the existence of black holes somewhere in the universe.

once claiming: 'I don't like it, and I'm sorry I ever had anything to do with it.' By a delicious irony, he created his famous thought experiment as a method of demonstrating the apparent absurdity of many quantum mechanical implications.

Yet many elements of QM have been tested and shown to work. Many of the more advanced ideas spring from the same well as those which have given us much of our modern technology – so how far can we go in accepting or denying the more imaginative theoretical offshoots? If we dismiss aspects of QM we are risking throwing out the baby with the bath water.

It may be that modern physicists have gone along a path that has led to ridiculous and quite wrong theoretical explanations for practical and workable phenomena. If this is ever shown to be the case, the parapsychologists will have scored a victory, but I suspect the answer to this lies some way in our own future, and to acquire it, we will probably have to wait.

References

1. Lyall Watson, *Supernature*, Sceptre, 1974, pp. 272–3.
2. Quoted in Hans Eysenck and Carl Sargent, *Explaining the Unexplained*, Weidenfeld & Nicolson, 1982, p. 13.
3. C.G. Jung, *Synchronicity: An Acausal Connecting Principle*, Routledge & Kegan Paul, London, 1972.
4. Lawrence Krauss, *The Physics of Star Trek*, HarperCollins, 1996, pp. 65–83.
5. Quoted in ref 2. p. 139.
6. John Gribbin, *In Search of Schrödinger's Cat*, Corgi, 1984, p. 229.

Chapter 7: The Agony and the Ecstasy

When you buy shoes measure your feet.

Chinese proverb[1]

Pain is a perfectly natural, even necessary physical process, but there is growing evidence that the human mind can train the body to negate pain and control a range of other bodily functions, bending the physical to the power of the will.

The incredible abilities of certain individuals, known in India as fakirs or yogis, have been documented since ancient times. Descriptions of fire-walkers, self-mutilators and tales of people capable of the most amazing feats of endurance are mentioned in the Bible, in the writings of the Ancient Greeks and in oriental texts. Today, displays of these talents may be witnessed by tourists around the world.

I have experienced such skills at very close quarters and in the most unlikely place – Venice Beach, Los Angeles. Amongst the robotic dancers, limbo artists, buskers and chainsaw jugglers, I spotted a small crowd gathered around a terrifyingly thin white American who was busy preparing a bed of glass made from a collection of bottles. He was joking with the crowd as he smashed the bottles and arranged the pieces on a sheet laid upon the ground and when he spotted me taking pictures I was called over and asked to help in his act.

Five minutes later, the glass heaped high, three inch jagged shards protruding from the pile, the performer, whose name he told me was Larry, had me steady a chair as he climbed onto the seat. A moment later he jumped feet first onto the pile of glass. The crowd fell silent, but Larry was totally unharmed. He then proceeded to leap up and down on the glass, rubbing his feet amongst the slivers as though he was paddling in a stream. When he had finished, he showed me his soles – there was no blood, not a single scratch.

In many cultures such acts of courage and control are treated as a religious ritual, a cleansing process and in ancient times facing such terrors was used as a form of judgement – the traditional trial by fire. According to one story, in 1062, the Bishop of Florence was accused of corruption by the saintly Peter Aldobrandini. Peter declared that he and the Bishop should both walk on fire if the clergyman wished to prove his innocence. A corridor of hot coals was prepared and a bonfire placed at each end. Peter Aldobrandini entered the flames, walked the path of flaming coals and through the fire at the far end totally untouched. When it came to the Bishop's turn, he wisely declined, opting instead to resign his position.

Such stories lead people to believe that these seemingly miraculous feats could only be achieved by the pure of heart, or through deep meditation and rigorous fasting, and today such beliefs are still common amongst those who perform the acts or watch agape. Modern research has shown that religious observance is only helpful because it imbues confidence. The ability to walk on fire, pierce the body with hooks or lie on a bed of nails unharmed is a question of applied science, training and belief in oneself.

One of the most dramatic examples of human endurance is the art of fire-walking. The Old Testament contains a vivid description of such an event in the tale of Nebuchadnezzar's attempt to execute Shadrach, Mesach and Abednego. According to the biblical account, the fire was so hot it killed the executioners taking the accused to the flames but left the victims miraculously untouched. 'The princes, governors and captains, and the king's counsellors, being gathered together, saw these men, upon whose bodies the fire had no power, nor was a hair of their heads singed, neither were their coats changed, nor the smell of fire had passed on them.'[2]

In isolation this could be treated as just another far-fetched Old Testament drama but the same, seemingly miraculous exhibitions are witnessed everyday and within recent years some of these have been subjected to serious scientific study.

In 1980 a team from the University of Tübingen in West Germany travelled to the annual festival of St Constantine at Langadhas in northern Greece to investigate the phenomenon. As the festivities reached a climax and the fire-walkers prepared themselves by chanting and performing ritualistic dances, the scientists set up their apparatus. They attached thermocouples to the walker's feet and electrodes linked to an EEG which measures brain-wave patterns were placed on the scalps of the yogi. The team noted the length of the fire pit at 4 m and that it was filled with coals and embers to a depth of 5 cm. The surface temperature was measured at 495 °C (932 °F).

After the display the investigators measured the temperature of the soles of the walker's feet to be 180 °C (356 °F), but could detect no sign of blistering or scorching. At the same time, the EEG recorded that the subjects had shown significantly increased theta rhythms during the walk.

At another ceremony in Sri Lanka, the psychic investigator, Carlo Fonseka visited several hundred fire-walking demonstrations where he took measurements of the fire-pit and monitored the walkers. He found that the fire paths were usually between 3 and 6m in length and 8 to 15 cm deep with fire temperatures lying between 300 °C and 450 °C (580 °F and 850 °F). But not contented with watching the performers in action, Fonseka decided to set up a laboratory test to quantify the ability of subjects to resist high temperatures.

The apparatus was simple; a 40-watt light bulb inside a metal cylinder. Volunteers were asked to place the soles of their feet on the top of the cylinder for as long as they could. The control subjects, people untrained in fire-walking felt some heat after between 6 and 10 seconds and after 30–40 seconds they experienced extreme pain. The fakirs from the fire-walking ceremony felt no heat for an average of 29 seconds and could keep their feet in position for up to 1 minute 15 seconds.

Fonseka's conclusion was that the fire-walkers all had soles with far thicker epidermal layers. They acquired this protective covering as a reaction to wearing shoes or sandals only rarely and it acted as a very efficient thermal insulation.

But this was not the only reason for their success as fire-walkers; the other crucial element, Fonseka claimed, was speed. He conducted a study of over 100 walks and found that the fakirs spent an average of only 3 seconds in contact with the coals during the entire walk, with each step requiring 0.3 seconds.

Even so, three seconds is plenty of time to receive severe burns and the secret, according to this set of results, seems to be a combination of conditioned soles and speed. This is borne out by frequent cases where ill-prepared walkers who insist upon 'having a go' end up in hospital with third-degree burns.

But this still cannot be the entire picture. There are several other factors to consider in accounting for success with the art of fire-walking, because it has been demonstrated safely by westerners with soft feet and with very little preparation. Hardened soles and a lightness of touch are obviously helpful, but some physiochemical aspects must be of equal importance.

Dr Jearl Walker, professor of Physics at Cleveland State University, has proposed something he calls the *Leidenfrost effect*. Johan Leidenfrost was the first person to notice that liquid exposed to sudden intense heat produces an insulating cushion of steam. Walker concluded that perspiration produced on the soles of the feet because of the excitement of the ceremony evaporates immediately it comes close to the hot coals. This creates a layer of protective steam on the soles of the runner which lasts long enough to cushion the impact of the heat.

Another, purely empirical, theory comes from physicist Bernard J. Leikind at U.C.L.A. in California who suggests that part of the explanation derives from the difference between temperature, heat and internal energy.

The words 'temperature' and 'heat' are often misused and it is easy to forget that temperature is merely a measure of the heat content, but heat is the result of molecular vibration within an object. Take for example the hot coal in the fire path. This is made largely of carbon which has a very regular chemical structure. But, unless a substance exists at absolute zero (–273 °C, 0K) its molecules will be in a state of constant vibration and the more internal energy they possess, the faster they vibrate. When one object warms up another it happens because some of the energy from the hot object has been transferred to the cooler one causing the molecules of the cooler object to vibrate faster.

Complications arise from the fact that different materials at the same temperature possess different amounts of energy. This seemingly odd result comes from the fact that different materials have different *specific heat capacities*, which signifies their relative ability to store heat. Perhaps more important still is the ability a substance has to convey or pass on its thermal energy. This is called the *thermal conductivity* of the substance, and the carbon in the coals of a fire path has a surprisingly low thermal conductivity. So, although the embers may measure a temperature of some 500 °C, this energy is not efficiently transferred to the walker.

To draw an analogy; imagine yourself walking barefoot across a collection of different surfaces on a hot summer day. Sand and metal would feel very much hotter than say rush matting or a wooden surface because the sand and the metal are better conductors and transfer their thermal energy to your feet with greater efficiency. This then is one very good reason why fakirs have not updated their display and replaced the hot coals with a length of heated metal!

Collectively, these effects go some way to explaining how fire-walking can be achieved without burning and how those who have prepared succeed over those who do not. But there remains one other requirement that in many cases can make a significant difference.

The power of positive thinking or self-confidence is not a supernatural quality, but merely a matter of applying instinctive abilities. The fact that the team of scientists from Tübingen found greatly increased theta rhythms in their subjects confirms the idea that fire-walkers are able to overcome pain by adopting a rarefied mental state. Theta rhythms are associated with meditation or deep relaxation, so it would appear that to prepare for a fire-walk or any other extreme feat like this, the yogi must channel his brain patterns and calm his body and mind.

Professor Leikind from U.C.L.A. and his colleague, William J. McCarthy, both ventured into the fire during their experiments into the physics of fire-walking and emerged unscathed. They do not have hardened soles and had no special training before the event, but are convinced that part of the reason they could do this was because they had prepared themselves mentally. They did their fire-walk at the home of Californian self-help prophet, Tony Robbins, who headed fee-paying courses instructing participants how to allow their theta rhythms to dominate their brain patterns.

Several techniques were employed to do this and were noted by the investigating scientists. The first trick was to ensure that the walk was scheduled for the early hours of the morning when people would normally be asleep. Staying up like this has the effect of depressing sensitivity. It also allows them to be more suggestible and to channel their theta rhythms with greater ease. Other pain-blocking techniques included special breathing patterns and chanting. Repeating such expressions as 'cool moss, cool moss' seems to help the participants to focus their thoughts, to concentrate on achieving their goal and blocking pain.

What surprised researchers initially was that these techniques did not just inhibit pain, but the majority of walkers showed no physical signs of heat. There was rarely a blister to be found amongst the devotees who enrolled on the course.

It may be that some participants have a greater talent for controlling their bodies than others. It is certainly possible that fakirs who have spent their lives training and are convinced of the religious aspect of the technique are able to manipulate some of the biochemical processes within their own bodies. This is known as biofeedback and a number of laboratory tests have shown that gifted or well-trained individuals can control such functions as heartbeat, skin sensitivity and muscle tension. For the majority of 'amateurs' who have managed to walk on fire after only a few hours preparation, self-confidence may allow a relatively painless journey, but the Leidenfrost effect, conductivity and speed have probably played the most significant role.

The ability to adopt a trance-like state induced by chanting and special breathing is actually a civilized version of a natural survival technique. Often prey fall into a cataleptic state when they are caught but not killed by a predator. Wildebeest often stand motionless with glazed eyes as a pride of lions bite off its legs and some tarantulas readily submit to the attacks of tarantula wasps.

The reason for this is twofold. Paradoxically, it offers the possibility of escape because, by not struggling, the predator is less likely to deliver a *coup de grâce*. The other, more relevant reason, is that by adopting a trance-like state in which theta rhythms dominate, the animal suffers less pain.

The stories of individuals slipping into an involuntary relaxed state and performing tasks they would otherwise consider unimaginable are surprisingly common. A friend of mine drove her car over a cliff and managed to crawl from the wreckage that had squashed her into a space a metre square. Despite sustaining a broken arm, severe concussion and facial injuries, she managed to walk half a mile to the nearest telephone, where she calmly called her father and waited for an ambulance to arrive. Today, she has almost no memory of the interval between escaping the wreckage and reaching hospital.

Part of my friend's ability to do this came from a flood of chemicals in her bloodstream. The hormone adrenaline constricts the blood vessels stemming blood flow and provides emergency energy supplies, also a mechanism known as the *extrinsic mechanism* comes into play under extreme conditions to help clot blood. But of key importance may be a natural ability to regulate the brainwave patterns and, in the case of extreme emergency, to create a balance of patterns that allow us to do what is necessary in order to reach safety.

In the late 1970s, Dr Wolfgang Larbig from the University of Tübingen carried out a series of experiments on an Indian yogi who was wired up to a collection of monitors and his brain responses compared to a group of control subjects.

The experimenters applied painful electric shocks to the fakir and the volunteers and monitored their heart-rate, skin conductivity and brainwave patterns. They found that after a short time the volunteers could take no more and looked exhausted, whereas the fakir had remained impassive throughout. They also discovered that the conductivity of his skin at the end of the tests was very different to all the other subjects, that his heart-rate was much slower and that he had produced strong theta rhythms throughout the experiment.

In order to do this, the yogi was thought to be practising a form of self-hypnosis and some psychologists are beginning to use this technique to help people overcome physical problems such as chronic pain or psychosomatic disorders. According to one enthusiast, the hypnotist Leslie LeCron: 'Essentially all hypnosis is self-hypnosis. The operator is merely a guide and the subject produces a result.'[3]

LeCron believes that the basics of self-hypnosis can be taught very easily, but the ease with which an individual can allow themselves to slip into a hypnotic state varies from person to person. The technique requires concentration on a simple fixed image (a candle is a favourite) which then allows the subject to focus their minds and channel their brainwave patterns into a slower rhythm. These are usually either alpha waves (normally associated with sleep and found at 8–14 Hz) or the ubiquitous theta rhythm at 4–7 Hz.

A similar practice is Transcendental Meditation. This became popular in the West during the late 1960s thanks to the Beatles' well-publicized flirtation with the art. At its simplest level TM is used by many people as a relaxation technique and others as a cure for insomnia, but for the real enthusiasts it is an avenue into some of the more advanced practices displayed by yogi and holy men. Central to the concept is the use of a simple word chanted silently over and over again. This is really a mechanism to send the brain into a relaxed state and to reinforce a single simple rhythm, a substitute for the candle or the simple visual image employed by hypnotists such as Dr LeCron.

TM was actually known about in the West long before the Fab Four popularized it. The nineteenth century English poet Lord Tennyson was a keen user, once describing it as:

> A kind of waking trance – this for lack of a better word – I have frequently had, quite up from boyhood, when I have been all alone. This has come upon me through repeating my name to myself silently, till all at once . . . individuality itself seemed to dissolve and fade away into boundless being, and this not a confused state but the clearest, the surest of the surest, utterly beyond words.[4]

Fire-walking is not the only form of demonstrating resistance to physical pain. A speciality of Sri Lankan yogis is hook-hanging. This involves hanging by half a dozen wires tied to hooks passed through the flesh of the back.

For many practitioners it is a form of religious observance and they see it as a way of atoning for their sins in much the same way that Catholics count rosary beads or pay penance. For the uninitiated it is extremely dangerous and even trained fakirs sometimes suffer infections as a result of their devotions.

Like fire-walking, the ability to hang by hooks is a combination of mental control and physics. The suppression of pain is again a result of controlling brainwave patterns using breathing techniques and accentuated by a self-belief in what they are doing. Physics plays a role in that the hooks have to be arranged in such a way that the fakir's body weight is distributed evenly.

When hook-hanging was first witnessed by western scientists

they were surprised to note that yogi rarely bleed from their wounds. To explain this we return to the extrinsic mechanism and the adrenaline rush experienced when an ordinary person is placed in an extraordinary situation. Like my friend's experience after her car crash, the body of the fakir responds to the self-inflicted crisis. The difference is that in the case of an accident, the response is automatic and involuntary, but in the case of the hook-hanger, the same physical responses can be turned on and off and controlled at will.

A famous image of pain endurance is the yogi laying on a bed of nails. Once more this is explained by a combination of mental preparation, physical training and science. Like the fire-walker and the hook-hanger, the yogi performs this act for religious reasons and so he is motivated to train, to be influenced by ritual and to allow positive thinking to dominate. This assists the production of enhanced theta-rhythm activity which can then be channelled to control a range of physical characteristics. As well as this, the fakir has been trained how to lie on the nails and how to lower himself and raise himself from the spikes so as to avoid penetration of the skin. Finally, the distribution of the nails is such that none of them has to bear too great a weight which lowers the risk of piercing.

Other yogis show a talent for piercing the body with spikes, wires and even rapiers, yet they rarely bleed despite rupturing blood vessels and tearing tissue. Training has enabled them to pierce themselves in specific places, carefully avoiding internal organs, major arteries and muscle. Those who perform this trick frequently have also produced areas consisting of unusual amounts of scar tissue which allows the wound to be re-opened more readily. But, aside from these precautions, the fact that the fakir can control bleeding is possibly down to a highly developed ability to monitor and manipulate many of the body's natural functions.

Although the vast majority of fakirs and yogi live and perform in India and the Far East, there are examples of westerners who have managed to acquire many of the skills of the holy man of ancient tradition. Probably the most famous of these was Harry Houdini, who enthralled audiences with his sensational act from the turn of the century until his death in 1926.

Houdini regularly escaped from strait-jackets in seconds, but he also survived being immersed in freezing rivers with his hands and feet manacled, claiming that he had defied death this way over 2000 times during his career. On one occasion he survived in a sealed coffin submerged in a swimming pool for an hour and a half, beating the record set by an Egyptian fakir. He claimed he could take any blow anywhere on his body as long as he had time to prepare. Ironically, he died from a ruptured appendix after a fan punched him in the stomach before he had time to tighten his abdominal muscles.

So what conclusions can the open-minded scientist draw from these acts of self-mutilation and pain resistance? Clearly, the mind can, to a degree, control the natural processes of the body, but just how far this ability could extend is difficult to judge. Some religious sects take the development of these powers to an extreme and followers devote their lives to honing their natural skills. There are apocryphal tales of holy men fasting for twenty years whilst sitting impassive in a full lotus. Such stories have never been substantiated and defy all the laws of science. But lesser demonstrations are quite common and could be attributable to a form of 'mind over matter' as well as sensible application of technique and sound science. What research and everyday experience highlights is that we all have the ability to come involuntarily through the most extreme conditions imaginable, which shows just how adaptable and responsive our bodies are. It is these natural talents the yogi exploits with dramatic and often startling results.

References

1. *Seven Hundred Chinese Proverbs*, translated by Henry H. Hart, Stanford University Press, 1937, p. 62.
2. Daniel 3,27.
3. Leslie LeCron, *Self-Hypnosis*, NAL, New York, 1964, p. 19.
4. Quoted in *Mind Over Matter*, edited by George Constable et al., Time-Life Books, 1988, p. 118.

Chapter 8: A Chance Of a Ghost

> From ghoulies and ghosties and long-leggety beasties
> And things that go bump in the night
> Good Lord deliver us!
>
> Anon.

According to one recent report, 1 in 6 people believe in ghosts and 1 in 14 claim to have seen one.[1] But, the problem is: What do we mean by the term? More than any other paranormal phenomenon, ghosts engender a range of emotions from terror to cold ridicule and to many they lie at the extreme edge of acceptable supernatural oddities. The primary reason for scepticism is that in all but a few rare cases, ghosts are thought by believers to be spirits of the dead. To accept this we must assume humans have souls and that the soul survives death. Beyond that we have to concede that the ghost has returned to the world of the living for a purpose which implies a strong link between this world and the next; for why else would a dead person care what happens here and now?

All of these steps in accepting the notion that ghosts are anything other than hallucination or tricks of the light, self-delusion or hoaxes, means we have to throw away every science book written during the past three hundred years. It also implies we as individuals are very much more important than everyday experience tells us we are and that we have personalities that survive physical death and then interact with the mundane dramas of earthly existence.

In spite of these criticisms, ghosts are perhaps the oldest and strongest held occult belief and have appeared in almost all cultures and all historical periods. According to their legends, the shamen of tribal cultures can act as conduits via which the living and the dead may communicate. The ancient Egyptians charted the progress of the individual after they had passed beyond this world in their *Book of the Dead* and in our own culture, the *Bible* is full of apparitions, ghosts and visitations. However, no one has so far produced irrefutable proof that ghosts are supernatural entities, nor can anybody provide a satisfactory theory linking apparitions with the spirits of those departed.

In view of this we must turn to science to give a reasoned and logical analysis of what ghosts are and how they have become so ingrained in our race consciousness.

There are two distinct phenomena that come under the umbrella term of 'ghost'; these are apparitions and poltergeists. Although, they may both flow from the same psychological spring, for our purposes they can be distinguished: apparitions are largely *passive*, whereas poltergeists are *active*. Apparitions are usually witnessed as images, which impart information or respond to living beings only rarely, whereas poltergeists are frequently reported to interact with the living and are often said to be malevolent, even murderous.

The first analytical study of ghosts was produced towards the end of the nineteenth century by a dedicated group of researchers of the paranormal calling themselves the Society for Psychical Research (SPR). The group formed around three scholars from Trinity College, Cambridge and endeavoured to investigate the current vogue for spiritualism using purely scientific means. Most active were Frederic Myers and Eleanor Sidgwick who wrote several academic books about paranormal phenomena and helped to conduct thousands of tests and investigations exposing fraud and hoaxes. By the end of the century the society had amassed 11 000 pages of reports later distilled into two important books, *Phantasms of the Living* and *Human Personality and the Survival of Bodily Death.*[2,3]

Early on, it was realized that apparitions could be fitted into three distinct categories. The first type are ghosts that appear to a single person and are thought by the 'sighter' to be a image of a dead person, perhaps a relative or a close friend. The second type is an image that appears to several people simultaneously. The third type (which may also occupy the second group) are visions of people who are still alive. This last variety are called crisis apparitions.

The majority of ghosts appear to individuals, invariably at night or in darkness and have little or no interaction with the environment. There have been countless cases of these 'simple apparitions', but absolutely no proof that such images are produced by a dead person deliberately trying to communicate with the living. Furthermore photographs taken of ghosts are invariably found to be fakes or to show quite natural aberrations of light or photographic anomalies.

Frederic Myers defined ghosts as: 'a manifestation of persistent personal energy, or as an indication that some kind of force is being exercised after death which is in some way connected with a person previously known on earth.'[4]

This is a purely empirical definition and has done little to satisfy those who believe ghosts are the spirits of the dead. What it suggests is that there may be some natural mechanism by which the energy of an event or a person could be recorded in the environment, and it is interesting to note that some cases of apparition show common characteristics. These might add weight to this idea without the need for supernatural causes.

The first thing one notices about ghosts is that they seem to be remarkably simple-minded things. As the writer, Colin Wilson has pointed out: '. . . a tendency to hang around places they know in life would appear to be the spirit-world's equivalent to feeble mindedness; . . . one feels they ought to have something better to do.'[5]

Simple apparitions usually perform a basic set of movements (and occasionally sounds) within a limited frame of reference. They appear to be tied in some way to a particular building or even a specific room, and to follow through repeated, identical movements – for example a walk along a corridor or across a room – and many appear only at certain times or under special conditions.

One example out of many thousands of such cases is that of the 'New Year's Eve Nun'. This ghost was first reported during the 1930s at a girls' school in Cheltenham, where on New Years Eve, a nurse and the headmaster both saw a nun in a sitting position at the edge of the school playground. There was no chair in the vicinity. The following New Year's Eve the pair saw the same woman in the same position.

Cynics have suggested there is no coincidence in the date except that perhaps the school staff had enjoyed an end of year tipple, but there are several interesting aspects to this and the majority of such cases that also match Frederick Myers' definition. This image did not interact with the modern environment at all and appeared to be quite unaware of the presence of the 'sighters'. The nun had been sitting on a chair in her own time, probably in a room that no longer existed and appears to have been some form of played-back image, a tiny, isolated segment of some previous event.

In one respect we see ghosts every day. If we watch a film such as *The Thirty-Nine Steps* made over sixty years ago in 1935, it is almost certain all the adult actors will now be dead (actually Robert Donat died in 1958 and Dame Peggy Ashcroft in 1991), yet we see them moving and talking, acting out their roles. It can be even more disturbing to see interviews of deceased celebrities, watching them talk about their plans and aspirations, laughing and joking with the interviewer.

Imagine how someone who had never heard of film or video would feel if they were shown a recording of someone they knew who had recently died. They could be forgiven for thinking they were seeing a ghost and might only accept it as a mechanically reproduced image if the process was explained to them.

Unfortunately we do not have a mechanism to explain how recordings of images could be made and played back without the use of a machine such as a television camera or a video system, but the idea has been around in one form or another since the early days of the SPR.

At the end of the last century it was postulated that some form of photograph of a scene or an individual could be taken and somehow projected at a future date. When cinema became popular it was a logical step to upgrade the technological comparison and suggest that a 'film' record was somehow made and played back. Today we think in terms of video, but how could such a process work?

All forms of recording rely upon an imprint being made in a specific medium. A photograph is produced when light activates chemicals to produce photochemical reactions that create a coloured product. In this way, the pattern of the original scene is transferred to the negative and reproduced on specially prepared paper. A musical recording is made by creating a series of very specific patterns in a plastic disc. When the stylus travels along these patterns or grooves the original sound is reproduced. As we improve our technology, the means by which this information transfer and retrieval is achieved changes but the basic principle is the same. A CD reads a pattern on a disc using a sensitive laser, a video player reads a pattern on a tape produced by fluctuations in the material of the tape sensitive to variations in an electromagnetic field. Parapsychologists suggest this principle of transfer and retrieval can also be applied to explain the appearance of ghosts.

However, there are two problems with this comparison. The first is there appears to be no device by which this could happen. The second is that in many cases, the preserved energy seems to be linked with human emotions.

Enthusiasts of this theory suggest a number of ways in which a recording device could be produced by the surroundings themselves. One idea is that buildings, (or rather the material used to build walls), contain chemicals that can act as receivers and holders of the information needed to produce a ghost image. Research is currently underway into the abilities of different materials to record images and it has been claimed that buildings containing unusual quantities of quartz are more common sites for the generation of apparitions.[6]

As well as considering the nature of the material acting as a storage and playback medium, we need to look at the energy needs of the system. A photograph is different to a film in that it is merely a static image. Ghosts invariably move, (even if within a very narrow frame of reference), so there must be something in the environment to reproduce a mechanical effect. When using a video player or even a hand-cranked cine projector, energy is expended in order to create a moving image. Walls do not usually move, so we would require some alternative method of producing an active image.

One possibility is that the sighter provides the energy. Alternatively, unusual atmospheric conditions could unlock a stored image. One common feature of ghost sightings is a sudden drop in temperature immediately before an apparition. It has been speculated that energy is taken from the environment in order to process and project the stored information. A further alternative is that several walls are involved in the projection in the same way a holographic image is constructed from a collection of superimposed two-dimensional recordings.

The second intriguing aspect of this system is the link with human emotion. One argument for the case that ghosts are action replays suggests that the initial image is produced within a specific region of space-time during a moment of intense emotional activity.

Support for this comes from the fact that apparitions often signify some intensely dramatic scene or traumatic personal event. One such case tells of a woman named Elizabeth Dempster who moved into a flat in London and immediately began to feel a brooding, unhappy presence. In an effort to dispel the morbid atmosphere, she decorated the flat in bright colours. It was then she began to see in the bedroom the image of a mournful woman dressed in Victorian clothes. After researching the history of the house she discovered that soon after it was built, it had been occupied by an Italian woman and her husband. When the woman learned that her husband had been killed suddenly, she locked herself in the bedroom and stayed there until she died of starvation and self-neglect.

Another example of an apparition thought by supporters to be linked with emotional impact are ghost battle scenes. These sometimes appear on specific days, perhaps when the environmental conditions are appropriate. Such spectral battles are often witnessed by several people simultaneously and can come complete with sound effects and even smells.

A famous report from the seventeenth century tells of a group of shepherds who claimed they witnessed a re-enactment of the Battle of Edgehill. This was one of the decisive conflicts of the English Civil War in 1643 when 500 men were brutally killed within a small area during a few hours of one another.

But how could emotions be recorded in a medium such as a building, or, in the case of a battle scene, by nothing but thin air?

Supporters resort to the concept that energy produced by the brain could be transferred and suggest that during particularly traumatic incidents, the human brain produces unusual brainwave patterns and is capable of projecting this energy.

Humans certainly do produce unusual brainwave patterns during moments of trauma or emotional anguish, but even during these times, the energy associated with such disturbance is many orders of magnitude too weak to have any impact upon the material world. However, it is perhaps possible that the combined energy from hundreds or thousands of simultaneous deaths or shared anguish could create a *gestalt* that imprints itself upon the environment.

In Hong Kong, the greatest number of ghosts are reported in buildings that were occupied during World War II and in particular those in which Chinese prisoners were tortured to death. The second highest population of ghosts in the city is to be found in the hospitals. But if humans were able to imprint emotion onto the environment, why is Hiroshima or Nagasaki not haunted by powerful after images of the thousands who died there simultaneously? Why is Belsen or Dachau not a region of space in which intense visible projections are seen on a daily basis? Could it be that the environmental conditions in those places were not right at the time to imprint images or that the conditions to play them back are not suitable today? One answer might be that in the case of the atomic bombing of Japan, the energy released during the explosions not only killed thousands instantly but stopped the mechanical process by which an image could be stored.

A variation upon this theme of recording and replaying images has been postulated by several modern-day researchers of the paranormal. 'Ghosthunter' Andrew Green who made headlines in 1996 after he was called in to find a ghost spotted several times in the Royal Albert Hall, claims ghosts are projections created by the sighter themselves.

Ghosts he believes are: '... forms of electromagnetic energy between 380 and 440 millimicrons of the infra red portion of the light spectrum ... If I was told my wife had been killed, this kind of information shocks me to the core. I picture her in my mind's eye. And that image is transferred to where I last saw her. It may be 50 miles away or upstairs in the bedroom, it doesn't make any difference, that image is there suspended till someone else picks it up.'[7]

So, by this reasoning ghosts are telepathic projections and we again come up against the problem of the energy needed to create such an image or to transfer the necessary information to another individual who sees it. There is also the fact that ghosts would only be seen if someone who had known the person was still alive themselves and able to create the projection.

If we reject the idea that ghosts are either recordings or projections we have to consider a limited range of other options, some of which might at first appear obvious. One of the original investigators of the paranormal, Eleanor Sidgwick of the SPR, produced a list of mundane explanations for ghosts that should always be considered before entertaining inexplicable sources. She listed these as: 1. hoaxing; 2. exaggeration or inadequate description; 3. illusion; 4. mistaken identity; 5. hallucination.

Hoaxing was rife at the end of the last century when the SPR was setting up in business. The craze for spiritualism had started earlier in the century and had become a booming cottage industry with an increasing number of mediums and spiritualists ready to fleece the gullible. It was an interest that crossed all social divides and enticed the superficial interest of artists, writers and curious scientists. Charles Darwin once attended a seance with other scientific colleagues, academics from Cambridge and the writer George Eliot, but came away from it even more sceptical than before.

In recent times there have been some famous hoaxes. 'The most haunted house in England', Borley Rectory in Essex, which burned to the ground in 1939 was supposed to be the site of some 5000 paranormal incidents during a period of a few years. It was made famous when the researcher into the paranormal, Harry Price, investigated the building and wrote a best-selling book about it called *The Most Haunted House in England* (published in 1940). During the 1950s the SPR investigated Price's claims and revealed that they were entirely faked.

Another example is the 'Amityville Horror' house in Amityville, New York which became the subject of a famous book and a Hollywood horror movie. The entire story was fabricated for purely commercial reasons by the one-time owners, George and Kathy Lutz and a lawyer, William Weber.

During the nineteenth century, there was money in spiritualism and mediumship just as there is today in successfully faking sensational paranormal experiences. People want to believe, they want to talk to their deceased loved ones and, in an age before the advent of television and virtual reality people were more impressionable, they would pay for the privilege. One of the functions of the academically minded SPR was to expose cheats and frauds and during the latter half of the last century there were even prosecutions and imprisonments as a result of exposés helped by the investigations of groups like the SPR.

Today hoaxes account for only a small percentage of ghost sightings and improved technology can usually spot obvious fakes, but exaggeration and inadequate description account for a significant percentage of apparitions. Often the witness is sincere but has misconceived an experience unintentionally, or else they have been subject to an optical illusion. Fear can distort, and illusions under stress are surprisingly common. If a lone witness sees what they think is a ghost in bad light or under unusual environmental conditions they can often be forgiven for misinterpreting what their eyes tell their brain.

However, the most interesting of all so-called *trivial* explanations for ghosts is hallucination and many researchers, both enthusiasts of the occult and empirically minded sceptics, accept that the vast majority of apparitions can be explained in this way.

Hallucination is an intensely researched psychological state that is surprisingly widespread. Back at the turn of the century one of the most useful pieces of data gathered by the SPR was a survey conducted on 17 000 people to determine the incidence of hallucination. They found that 2300 of those asked had experienced an hallucination sometime during their lives, and according to a modern day survey of American college students, 70% claimed they had experienced the auditory hallucination of hearing voices whilst awake.[8,9]

In his book, *Fire In the Mind*, the psychologist Ronald K. Siegel has said of hallucinations:

> In the past, hallucinations were often regarded as the exclusive domain of the insane. Through the research and cases in this book, we begin to understand that anyone can have them. They arise from common structures in the brain and nervous system, common biological experiences, and common reactions of the brain to stimulation or deprivation. The resultant images may be bizarre, but they are not necessarily crazy. They are simply based on stored images in our brains. Like a mirage that shows a magnificent city on a desolate

expanse of ocean or desert, the images of hallucinations are actually reflected images of real objects located elsewhere.[10]

The most common time to see a ghost is late at night and usually at the point of going to sleep. The 'ghost at the end of the bed' is the stuff of legend and the main stay of Hammer horror films yet there is a well understood reason for this. As the body switches from the voluntary nervous system, (the system that allows us to function in our everyday lives) to the involuntary nervous system, we commonly experience what are called *hypnagogic hallucinations*. One interpretation of these is that 'our wires get crossed'; the brain is momentarily confused by the switch from one nervous system to the other and images are dredged up from either deep in the conscious memory or from the subconscious. This, it is believed, accounts for the vast majority of apparitions.

These visions or hallucinations can seem very real and may be accompanied by auditory sensations or even smells. A similar experience is sleepwalking. Many people have at one time or another had the odd experience of suddenly coming to in the bathroom or sitting in front of a blank television screen in the sitting room. Often these experiences seem very real at the time but are almost totally forgotten by the following morning.

Another common form of sleep-related hallucination is what has been dubbed *hypnopompic hallucination* and occurs when we awaken. Again, this is due to the body switching nervous systems, this time from the involuntary to the voluntary, and may account for a large number of cases of apparition.

Both hypnagogic and hypnopompic hallucination may also be used to explain what have become known as 'hitchhiker apparitions'. Since the beginning of the car-age, an increasingly common phenomenon within ghost mythology describes accounts of drivers seeing ghostly figures in or at the side of the road. These sightings often occur along stretches of road famous for particularly grisly accidents or known in the region for apparent spectral activity. Sometimes drivers have even reported knocking down people, feeling the bump of the body under the car and when they have gone to see what had happened they are left staring at empty tarmac.

Even more dramatic incidents tell of drivers tending someone they believed they had hit, covering them with a blanket only to find the body had vanished by the time the police arrived. In 1979 a driver named Roy Fulton claimed he picked up a male hitchhiker along a stretch of road in Stanbridge, Bedfordshire late at night. The young man opened the door and sat silently on the back seat ignoring Fulton's attempts to begin a conversation. Only when he turned to offer the hitchhiker a cigarette did Fulton realize the boy had vanished.

Such incidents are open to ridicule and to claims of hoaxing and perhaps a large proportion of them are deliberate frauds, but in some cases they could be put down to either hypnagogic or hypnopompic hallucination. Drivers sometimes fall asleep at the wheel and it is quite possible hallucinations could occur as they lose consciousness or wake suddenly. These brain-generated visions are then amplified by environmental effects and individual circumstances. Driving alone along narrow country roads in the dark can induce suggestive images in the mind and speeding along a seemingly endless stretch of featureless motorway is often almost mesmeric.

A related phenomenon is crisis apparition, where sighters sense the presence of someone they know either close to or at the point of death. There are many documented cases in which people have apparently seen projections of close relatives or friends who were in a crisis situation at the time, (often immediately before their moment of death). In their book *Phantasms of the Living*, Edmund Gurney, Frederick Myers and Frank Podmore documented 701 cases of apparent crisis projection and admitted they could not explain many of these incidents.

One of the most famous stories of crisis apparition comes from the 1930s. One stormy, freezing night in the mid-Atlantic, a one-eyed English pilot named Hinchliffe was attempting the first east-west crossing of the ocean with his female co-pilot when suddenly their biplane hit bad weather. The high winds tossed it around, the compass was disturbed by magnetic interference and without a reference point for hundreds of miles in any direction, they were soon hopelessly lost. The plane began to nosedive towards the waves, its engine screaming in protest and a moment later it hit the water killing pilot and co-pilot instantly.

The same night, two friends of Hinchliffe's, Squadron Leader Rivers Oldmeadow and Colonel Henderson were steaming towards

New York aboard an ocean liner several hundred miles away from the scene of the crash. Neither of them had seen Hinchliffe for some time and they were totally unaware he and his female colleague had attempted the flight. It was in the middle of the night, just when, according to later corroboration, Hinchliffe's plane hit the storm, that Colonel Henderson, dressed in his pyjamas, burst into his friend's room shouting. 'God, Rivers, something ghastly has happened. Hinch has just been in my cabin. Eye patch and all. It was ghastly. He kept repeating over and over again, "Hendy, what am I going to do? What am I going to do? I've got the woman with me and I'm lost. I'm lost." Then he disappeared in front of my eyes. Just disappeared.'

Supporters of paranormal explanations for ghosts and apparitions have proposed that this incident and others like it are due to a form of emergency or crisis telepathy, that at the moment of death the human brain is capable of transmitting an image or news of their situation, perhaps as a final survival attempt. But it is also possible to see such events in a far more prosaic light.

Firstly, there is the strong possibility of hallucination. The sighter in this story, Colonel Henderson, may well have enjoyed a pleasant evening at the captain's table before turning in for the night and could have experienced an alcohol-induced hypnagogic hallucination.

But, say the enthusiasts, how does this account for the fact that Henderson had no idea his friend was in the middle of a risky flight?

The answer is; he almost certainly did know about it subconsciously. It is possible he had been reading a newspaper and noted subliminally an article about his friend attempting an Atlantic crossing without reading the piece or even realizing consciously that he had spotted it. This could then have enhanced his hallucination providing the subconscious image around which he produced the vision.

An alternative suggestion for this and many other cases of crisis apparition is that people have a subconscious knowledge of an event but need to create an hallucination in order to process the information through their conscious mind. The most usual reasons for this inhibition are fear and guilt. These emotions could force the conscious mind away from analysing or thinking about a situation and the brain then has to resort to filtering the information through an alternative system where it does not meet the same resistance.

A final explanation for hallucinations appearing as crisis apparitions is wishful thinking or comfort thinking. When this is the source, the visions are called *need-based hallucinations*.

Most people hope there is an afterlife and many, especially those who feel insecure or fearful, can experience this desire so strongly they produce 'evidence' to support their wishes. To them, a ghost is proof of an afterlife and so their subconscious mind is empowered to conjure up an appropriate image. In other situations people can imagine they are being visited by a comforting, supportive figure who either warns them of imminent danger or gives them extra impetus to fulfil a difficult task. The record-breaking driver Donald Campbell claimed he had been visited by his father Sir Malcolm Campbell on many occasions and believed he had been sent to warn him of impending danger.

Of a quite different order to apparitions is the phenomenon of haunting, especially poltergeist activity. Hauntings are usually witnessed by several people and encompass a wide range of apparently supernatural activities – materializing and dematerializing of objects, noises, smells, and on rare occasions violent, even life-threatening incidents.

When a group of people all witness the same set of experiences it is very difficult to explain them as hallucination, hoax or any other natural process. But many do yield to rather mundane causes if the investigators probe deeply enough.

The first stage in any investigation of a haunting is to eliminate natural sounds and smells. These may take some searching and enthusiasts of occult explanations are fond of listing cases in which months of investigation into particular hauntings have been wasted and no link to natural causes detected. These cases are rare and do not offer proof of supernatural activity, they simply show the researchers did not investigate thoroughly enough.

After hoaxes, natural causes and illusions have been ruled out, we can again turn to hallucination. Surprising as it may seem, it is possible for a group of individuals to experience the same induced images. This phenomenon is called *mass hallucination* and is brought about when one of the group is a stronger personality than the others and creates a convincing suggestion which is then adopted by the others. This explains many cases of hauntings involving parents and children where one of the adults (or in rare cases, one of the children) unwittingly implants the idea, after which fear and anxiety take over.

Enthusiasts of the occult take this idea and add an element of the supernatural to create what they call the *infectious hallucination theory*. This proposes that one of the group experiences an hallucination which is then transmitted telepathically to the others. But psychologists have shown through experiment that this mechanism is actually quite unnecessary. If the correct blend of personalities are put together in an atmosphere of perceived danger and fear an hallucination created by one of the stronger personalities is infectious enough to spread to the others without the need for telepathy.

The final explanation for poltergeist, and one supported by some enthusiasts of the occult, is the idea that emotional disturbance in human beings can be projected into dramatic physical events. The theory suggests that people with the ability to project telepathic images may, if they are placed in an emotionally challenging situation, produce enough psychic energy to move furniture or throw objects across a room.

As we saw in Chapter 4, the energy needed to do this is of a quite different order to that generated by the human brain and is, according to the known laws of physics, completely impossible. A less problematic explanation may be that a human source of hallucination is planting images into the minds of the witnesses. Again there is no need for telepathy in this situation. If the creator of the image is a strong enough character, they may be able to induce hallucinations in the other witnesses. They may even be able to convince those on the receiving end of any violent activity that they actually feel pain or have cuts, burns and bruises.

Poltergeist activity has been shown to centre around children and teenagers more frequently than adults; in particular, pubescent girls appear to be a common source. This has lead occultists to the idea that hormonal and emotion imbalance enhances PK abilities in these young women. A more logical explanation might be that during a time in which body and brain chemistry is in a disturbed state, the subjects could be capable of encouraging hallucinations in those around them by suggestion and emotional manipulation. Mothers placed in highly stressful situations thanks to the growing pains of their teenage girls might be particularly susceptible, especially if the notion their home is haunted has already been 'seeded' in their minds.

For at least a century believers in the paranormal source of ghosts and apparitions have tried to provide clear evidence for their claims. During the nineteenth century, the fashion for mediums and spiritualism succeeded in convincing large numbers of people that ghosts are supernatural, and it was not until sceptics exposed many practitioners as frauds that the popularity of seances diminished.

Photographic evidence is also very weak. Most of the examples that have made their way into popular culture show what look like pantomime ghosts with sheets draped over their spectral heads. Those that do not show cavaliers in full regalia or bug-eyed demonic creatures illustrate what look like faults on the negatives – bursts of light or optical aberrations. As far as I am aware there is no convincing film or video footage of ghosts or apparitions. If ghosts are visitors from an afterlife and are as common as people believe, then this is rather surprising.

One striking development which has been claimed as evidence for the supernatural origin of ghosts is *electronic voice phenomenon*, EVP. This is the name for the process where background recordings apparently throw up the sound of a dead person speaking.

This phenomenon first came to public attention in 1920 when the October issue of *Scientific American* carried a feature about the famous inventor, Thomas Edison. In the piece, Edison claimed he was working on a device that could be used to communicate with the dead. Most of his contemporaries believed he had finally gone senile and his ideas were ignored by the scientific establishment. Not surprisingly, by the time of his death in 1931, he had still failed to deliver the promised machine, but a few months earlier, an American psychic named Attila von Szalay claimed he had heard the voice of his deceased son and later 'trapped' these voices using a 78 rpm record cutter. Three decades later, in 1959, a Swedish opera singer named Fredrich Jurgenson was listening back to tape recordings of birdsong when he thought he heard human voices speaking in Norwegian far back in the sound mix.

Soon, others were following the lead of Edison, von Szalay and Jurgenson and books about the phenomenon began to appear. The most famous was *Breakthrough*, published in 1971 which documented over 70 000 recordings, many of which were made by tuning a radio between stations and recording the resultant white noise.

It is difficult for the scientist to accept that these recordings carry messages from the spirit world. The first and most obvious criticism is that the interpretation of the recordings is totally subjective. Different listeners hear different things on the tapes. Only when a stronger personality insists a certain voice is talking in a specific language and delivering what they conceive to be a sensible message, do others begin to hear the same sentences.

Second is the fact that human beings seem to have an in-built inclination (in many cases, a need) to find order within chaos. This is thought to be a deep-rooted survival tool, an ability to analyse and find patterns in seemingly random events. Such a skill enables an individual to be better prepared for unpredictable outcomes – a form of psychological preparation. It also demonstrates an instinctive need for security: order equates to stability, which means safety. Those who hear distinct voices amidst the random signals and recordings are deluding themselves through a deep psychological need.

A third explanation for this effect has nothing to do with the human subjects but is based upon a fundamental universal law – chaos theory.

A chaotic system is defined as one that shows 'sensitivity to initial conditions'. What this means is that a slight variation in the initial state of identical processes can lead to very different end results. For example, a speck of dust floating on the surface of a pair of oscillating whirlpools displays chaotic behaviour. The particle appears to move randomly and its course becomes increasingly unpredictable, but the way it moves during one experiment will also be very different in another depending on the initial conditions of the test.

Chaos is observable in everyday situations. The world's weather systems are chaotic, which makes forecasting notoriously difficult.

A dripping tap behaves chaotically, and it is even suspected that global financial systems show chaotic behaviour which means long term speculation is riskier than many investors realize.

The concept of chaos can be visualized by a very neat example which has become known as the *Butterfly Effect*. This says that the flutter of a butterfly's wings in England could eventually create a thunderstorm in Australia. Although this would appear to be fantastic it is possible because the tiny disturbance created by the butterfly is an example of altering the initial conditions. Because weather systems are sensitive to initial conditions, this minuscule effect can become amplified so that via a long and complex chain of events it can be seen to produce a thunderstorm the other side of the world.

So what has this to do with 'ghost voices' captured on tape?

One of the striking characteristics of chaotic systems is that within chaos there is also order. This, scientists believe, is how an apparently chaotic universe throws up 'islands' of order. Our earth, our civilization, us as individuals, could all be short-lived, local anomalies of relative order in an ocean of randomly shifting particles we know as the observable universe. As James Gleick says in his international bestseller, *Chaos*: 'The simplest systems are now seen to create extraordinarily difficult problems of predictability. Yet order arises spontaneously in those systems.'[11]

White noise from a tuned-out radio or background sounds from a recording of birdsong can readily create a chaotic system. But within that chaos there may be elements of order, shifting, fleeting fragments of organization. When an enthusiast, a believer in the supernatural source of ghosts listens to these patterns they may pick out momentary parcels of order and amplify them into coherent, but nevertheless, imaginary sentences.

Ghosts certainly exist, but there is no hard evidence to suggest any form of personality or soul survives the death of the brain. Therefore ghosts appear to derive from other sources. These sources are plentiful and varied, ranging from deliberate fraud to mass hallucination. It may also be possible that a form of energy representing the physical characteristics of a person could be trapped by an unusual confluence of environmental conditions. It seems to me unnecessary to add to this the idea that some form of psychic energy created by extreme emotion is also trapped. Many ghosts proven not to be fakes, illusions, hallucinations or freak environmental effects appear to be limited in their movements and merely re-enact a specific scene endlessly. Occasionally, the sighter feels depressed by the apparition or is sensitive to an oppressive mood, but this is almost certainly an effect created by the mind of the individual, not the ghost. If such a record and playback system exists, it is a purely physical phenomenon, perhaps generated initially by extreme emotion, but one which science has so far failed to quantify.

Until physicists and biologists find a way to study the possibility that apparitions are replayed images from the past, we can only assume ghosts derive from the minds of those who see them, projections of our own desires and fears, and that they have no material form in the physical world.

References

1. G. Gorer, *Explaining English Character*, London, Cresset, 1955.
2. Gurney, Myers and Podmore, *Phantasms of the Living*, Kegan, Paul, Trench, Trubner and Co., 1918.
3. Frederick Myers, *Human Personality and its Survival of Bodily Death*, Longmans, Green and Co., 1927.
4. Ibid.
5. Colin Wilson, *The Occult*, London, Hodder and Stoughton, 1971.
6. Jenny Randles and Peter Hough, *Encyclopaedia of the Unexplained*, Michael O'Mara Ltd, 1995, p. 141.
7. Andrew Davidson, 'The Spectre Inspector', *The Sunday Telegraph*, May 5, 1996, p. 3.
8. H. Sidgwick, E. Sidgwick and A. Johnson, 'A Report on the Census of Hallucinations', *Proceedings of the Society of Psychical Research*, 10, 25–422, 1894.
9. Terence Hines, *Pseudoscience and the Paranormal*, Prometheus Books, New York, 1988, p. 61.

10. Ronald K. Siegel, *Fire in the Brain: Clinical Tales of Hallucination*, Plume, New York, 1993, p. 11.
11. James Gleick, *Chaos*, Cardinal, 1988, pp. 7–8.

Chapter 9: The Lost and the Lonely

> But first on earth, as vampire sent.
> Thy corse shall from its tomb be rent.
> Then ghastly haunt thy native place
> And suck the blood of all thy race.
>
> Lord Byron

Creatures of the night, blood-sucking beasts and sea dragons haunt all of us subconsciously. Such monsters stalk our nightmares and play on our imaginations: they are universals, race images, Jungian archetypes. Some may also have their roots in scientific reality.

The umbrella term *mythical beasts* has been used to describe a vast collection of creatures from all parts of the world, creatures that have appeared in the history books of almost all cultures and wander the pages of literature and art. But recent investigations have begun to show that the term *mythical* may not be so appropriate as some sceptics believe. These animals are certainly shy in the extreme, yet collectively, the witnessed cases, photographs and filmed archives show that our world is almost certainly home to a host of creatures about which we have only begun to speculate. These could range from isolated remnants of prehistoric creatures to rare genetic aberrations.

There seem to be two distinct groups of creature that have not yet been fitted into any genus accepted by science. I'll call these *human variants* and *evolutionary cul-de-sacs*. The first of these are all creatures which approximate to human beings, but possess extraordinary characteristics and include vampires, mermaids and werewolves. The second category consist of beasts which would appear to have evolved outside the avenues observed by modern biology, creatures such as sea monsters, the Loch Ness monster, the yeti and the bigfoot.

Of human variants the most intriguing and the most frightening is the vampire. According to the theologian Heinrich Zopfius writing during the early eighteenth century:

> Vampires issue forth from their graves in the night, attack people sleeping quietly in their beds, suck out all their blood from their bodies, and destroy them. They beset men, women, and children alike, sparing neither age nor sex. Those who are under the malignity of their influence complain of suffocation and a total deficiency of spirits after which they soon expire.

In Europe, vampires have been known about since at least the early Middle Ages and have persisted in the modern imagination partly because of the abiding fascination they have held for writers. The most famous vampire in literature is the eponymous anti-hero of Bram Stoker's *Dracula*, written in 1897. Although the public image of a vampire derives largely from this creation, by the time the book was published, vampires had already become a mainstay of Gothic horror literature, a genre popular since the eighteenth century. *The Monk* by Matthew Lewis written in 1796 was one of the original vampire novels and a massive work called *Varney The Vampire* was published in the 1840s and ran to over 1000 pages.

All of these books convey the image of vampires as blood-sucking humanoids who cannot endure sunlight and prey on their victims in order to survive. They all have a vulnerable aspect to their characters and convey a heavy sense of eroticism. Since the eighteenth century when these stories first became popular the image of the vampire has remained remarkably consistent and appears to be an almost timeless image of corruption, power and sadness.

Amazingly, there are today a growing number of people who claim to be vampires. The films *Interview with a Vampire*, the 1990s remake of *Dracula* and Tarantino's *From Dusk Till Dawn* have all helped to encourage a neo-gothic subculture that appears to have inspired some fragile mentalities to buy themselves coffins and shun the sunshine, (most especially the lemon glow of Los Angeles).

The Internet increasingly plays host to a surprisingly healthy community of vampires and the confessions of blood-obsessives have appeared in several recent books, including *Chaos International* in 1992 and Rosemary Ellen Guily's *Vampires Among Us*.

True-life stories of modern-day vampires are becoming almost commonplace within certain communities. Jack Dean, a self-confessed human blood drinker, developed his taste for the stuff after he was injured in a car crash and another, Philip Hine, cultivated his fetish when he was sprayed with blood after a friend cut themselves badly.

According to one report, there are currently 36 registered human blood drinkers in Los Angeles and at the last count 700 Americans claimed to be vampires. The fact that none of these are likely to be the genuine article but are victims of an obsession or induced fixation makes the craze no less alarming. The *X-Files* episode '3' delved into this chilling underworld and came closer to actuality than almost any other episode so far broadcast. But at a time when Aids is rife, the fringe club scene in which 'love nips' are exchanged like kisses may be seen as a telling sign of the level of psychological damage some people live with, especially at the frenzied heart of the end-of-millennium metropolis.

Yet in spite of all the hysteria and sensationalism surrounding the concept of vampires, it is now believed that some vampiric characteristics have sound medical foundations and that the legend derives from a series of genuine genetic anomalies.

The mystery of the vampire probably began in Dark Age Bohemia. In those times small communities often lived in complete isolation and entire populations could lead their lives without every travelling beyond the valley in which their village or town was located. Quite ignorant of a larger world beyond the encircling mountains, such communities were forced to in-breed which eventually caused what biologists refer to as a lack of *genetic diversity*.

Evolutionary biologists believe the purpose of sexual reproduction is not simply to ensure propagation of the species, but to allow a healthy mixing of genetic material throughout the population. This is why the taboo of incest was created in primitive societies: even then it was recognized that mating with close relatives was likely to end in a high incidence of natural abortions or deformity. In a broader sense, lack of genetic diversity weakens a species and at the very least, genetic flaws begin to crop up, some of which can be serious. Unless fresh genetic material is introduced from time to time, genetic faults become increasingly common and can grow more dangerous with each generation.

According to a theory proposed by David Dolphin in 1985, amongst many defects, an hereditary disease called porphyria could have developed through lack of genetic diversity in some isolated villages of Bohemia. This disease prevents the production of a protein that is responsible for binding a chemical called a porphyrin ring to iron found in the haemoglobin of the blood.

The consequences of this are wide-ranging and could help to explain many of the characteristics usually associated with vampires. Firstly, those suffering from porphyria look severely anaemic because their haemoglobin is not being utilized efficiently and their blood is not as oxygenated as it should be. Secondly, the porphyrin rings that cannot do their job are deposited in the subcutaneous fat beneath the skin. The chemical is photosensitive and in sunlight it can release electrons which damage the skin and may cause severe blistering – hence the vampire's fear of sunlight.

Legend tells us that vampires cannot stand garlic and there may be a sound biochemical reason for this too. Garlic contains enzymes that might, under the correct conditions, replace the function of the missing protein in porphyria sufferers. If this is consumed by a vampire, they receive a sudden rush of the necessary biochemical which could kill them.

A further complication caused by the malfunction of the porphyrin ring is that gums recede. This has the effect of making the sufferer's teeth look larger than normal and could be the source of the image of vampire's fangs. Furthermore, the inability to bind the porphyrin rings could create a craving for blood in the genetically defective individual which might account for the blood-lust at the heart of the vampire legend.

Comparisons and links between fact and myth could extend even further. According to accounts in popular literature, vampires are often associated with aristocrats or noblemen. This may have its roots in the fact that the deficiency of the protein causing the blood disorder often only appears after the sufferer reaches sexual maturity (most usually around the age of 16 or 17). This also happens to be the age at which females in such communities were married. The link arises from the fact that in the Bohemian community where the myth could have originated, the chief or leader would have taken a young bride, possibly a virgin, and soon after she was lodged in his castle she would grow pale and suffer the effects of porphyria. To the simple villagers this would be clear proof the feudal Lord was a blood-sucking creature and the tale became elaborated and increasingly far-fetched as it was passed on from generation to generation. In fact it was merely a coincidence.

Similar genetic anomalies could also explain other forms of human variant. On some Pacific islands, the local diet consists largely of shark and dugongs (a herbivorous marine creature sometimes called a sea cow). Both of these creatures have livers extremely rich in vitamin A. This vitamin is most commonly found in a form called retinol which has been found to be the cause of certain rare birth defects. It is quite possible that at some point in the past children with webbed feet were born on these islands. If some of these cases were bad enough, the tale could have crept into folk legends and eventually entered western mythology via tales communicated to early explorers – hence, the mermaid.

Genetic defects might also explain the legend of the werewolf. A boy called Jean Grenier who lived in the early part of the seventeenth century was born with a deformed jaw that endowed him with a canine appearance. He also appears to have been mentally retarded and was caught attacking a shepherdess. He later confessed to a series of murders and from records appears to have believed he was some sort of human beast or werewolf.

Tales of man-beasts, men who were part wolf or part bear were common in the seventeenth and eighteenth centuries. Many of these seem to be linked with mental illness and it is interesting that central to the legend of the werewolf is the notion that transformation usually occurs at the time of a full moon. For a long time, the moon has been associated with mood swings in humans and although there has never been a clear empirical link between the movement of the moon and the brain state of people, many enthusiasts insist there is one. It has been speculated that the gravitational field of the Earth's closest neighbour can engage in some form of interaction with the electronic currents within the brain and affect mood in the way the close approach of a thunderstorm is known to cause depression in rare, sensitive individuals.

In an alternative theory, occultists point to the fact that the moon exerts a powerful mechanical (gravitational) influence upon the Earth resulting in the production of tides. Perhaps, they speculate, the moon could also alter the fluids in our own bodies. This could in turn produce biochemical affects upon our brains which then alter our emotions and our moods.

The position of the moon relative to the Earth does affect the gravitational field of the planet and might conceivably facilitate a tiny variation in certain physical characteristics within our brains. These could include changes in the water content or distribution in the cells which might therefore impose some shift in fundamental chemical processes that play a role in determining mood. If, through some unknown mental aberration or defect, an individual is highly sensitive to this theoretical effect, it could play a role in altering their psychological make-up, perhaps pushing them into psychotic, even murderous behaviour. If they are also physically deformed as Jean Grenier appears to have been, gullible people may view the individual as some form of dangerous freak, part man, part beast.

A famous case is that of Peter Stubb who lived in Germany in the sixteenth century. He was said to have made a pact with the devil who gave him a special wolf fur belt that transformed him into a werewolf. During the course of a decade, he killed several people including young children and pregnant women and was finally caught after he savagely murdered his own son. He was tortured on the wheel and confessed to a multitude of crimes before being decapitated and burned.

Looking at more mundane explanations, the legend of the werewolf could simply be a concoction based upon fear and exaggeration. All primitive races feared wild animals and the wolf was probably the most ferocious animal to be found in Europe in modern times. It is not surprising that legends and myths grew up around them.

An alternative theory is that werewolves were nothing more supernatural than rabies sufferers. Rabies was a common disease until this century and was responsible for the agonizing death of many unfortunates. The biologist Louis Pasteur witnessed the suffering of victims and regularly sanctioned euthanasia. The notion that humans could transform into wild beasts could have derived from the very real and horrifying sight of someone who had been bitten by a wild dog suffering the final agonies of rabies.

During the twentieth century stories of human variants have become increasingly confined to the pages of Gothic horror novels and Hammer horror films (unless, that is, you visit the clubs of Los Angeles!). But, the sightings of creatures that fit under the sub-heading of evolutionary cul-de-sacs or ECSs, have increased rather than diminished. Today, ECSs ranging from sea beasts to yeti are reported from all parts of the world. And along with these sightings has come a growing collection of photographic and video evidence.

Probably the most famous example of what believers claim to be an ECS is the Loch Ness monster or 'Nessie'.

Loch Ness is the largest and deepest fresh water lake in Britain and is some 37 km (23 ml) long and up to 230 m (750 ft) deep. It was formed around 250 million years ago but as a result of the most recent Ice Age, it was only cut off from the sea in relatively recent times – a little under 7500 years ago. Peat on the bed of the Loch makes the water unusually dark and with 3-metre high waves not uncommon, many see it as a small inland ocean rather than a lake.

The myth of the monster began in the sixth century when St Columba is said to have seen the beast of the Loch and there have been reports of sightings ever since. Nessie first hit the headlines in 1934 when a gynaecologist Robert Wilson claimed to have seen a large marine creature rising its head and neck out of the water and took a photograph which was soon splashed across the front pages of newspapers around the world. Since then tens of thousands of visitors flock to the Loch every year hoping for a glimpse of the monster and many claim to have seen something odd in the murky waters.

Robert Wilson's photograph, which became known as the surgeon's picture, was eventually exposed as a fake when the perpetrators owned up that the 'monster' had actually been part of an elaborate prank. Apparently, the beast in the picture had been made from a clockwork toy submarine and some canvas sheeting. Nevertheless, such revelations have done nothing to deter increasingly sophisticated searches of the Loch.

During the 1970s there were two major searches sponsored by the Academy of Applied Science from Boston, one in 1972 and the other in 1975. These involved taking underwater photographs and sonar scans of the Loch. The photographs from one of the searches revealed the upper torso and the 'gargoyle-shaped' head of a monster. Sadly, later analysis showed the team had photographed a giant plastic model of a monster used in 1969 during the filming of *The Private Lives of Sherlock Holmes.*

Undeterred, another team returned to Loch Ness to conduct Operation Deepscan in 1987. This ambitious project involved a fleet of 22 boats creating a sonar sweep the entire 37 km length of the Loch. Again, the search provided no definitive results or clear photographs, merely a few blurred frames that were later revealed to show nothing more exotic than tree trunks and assorted detritus. The other problem with this search was that the sonar did not reach to the bottom of the Loch and could cover no more than 80% of the volume of water. It is easy to imagine any shy creature living there heading for the bottom as soon as a flotilla of boats passed across the surface, so it may have been missed anyway.

But just because we have not yet seen the monster close up or found evidence it exists does not mean it is definitely not there. If Nessie does exist, what could it be and how has it survived?

The most popular suggestion is that the Loch Ness monster is not one creature but a community of plesiosaurs. The plesiosaur was an aquatic reptile that ranged through the oceans of the world from the late Triassic until the end of the Cretaceous periods about 195 million to 65 million years ago. The theory is that a group of plesiosaurs survived the extinction of the dinosaurs at the end of the Cretaceous period. Then, much later, a community became trapped in the Loch when it was sealed off from the sea and have lived there ever since.

Although there are significant problems with this theory it is by no means an impossibility. The central idea that small groups of creatures could survive species-extinction is actually not so far-fetched as we might imagine. In 1938, African fishermen found an armour-plated fish two metres long in the Indian Ocean. After careful analysis it was found to be a coelacanth, a creature thought to have become extinct 70 million years ago. This specimen had evolved into a much larger fish than its prehistoric forebears and is thought to be part of a community living deep in the ocean.

Evolution is the process by which species develop utilising two factors – natural selection and random chance. If a characteristic appears in an individual, it may be passed on to future generations if that individual survives long enough to breed. Both good and bad characteristics are passed on to future generations but species evolve into improved or better-adapted forms because 'good' characteristics make individuals better able to survive, stronger and dominant and therefore more likely to pass on their design improvement. Luck comes into the equation because changes in genetic characteristics, or mutations, appear randomly.

All species evolve, but the environment can affect the rate of evolution. If the environmental conditions are stable for a long period and the community can survive within narrow limitations, evolution will probably be slower than it would in a more challenging, competitive environment. It is possible that a small community of creatures such as a group of plesiosaurs would have evolved far slower than other species throughout the world during the same time period – say 65 million years.

But the problem with isolation, and a major limitation to the survival chances of small communities, is the damaging effect of restricting the genetic diversity of the species. A small community of plesiosaurs is no different to an isolated Bohemian community of humans in that genetic faults could develop and threaten the health and eventual survival of the group. If we presume that the original community consisted of no more than a few dozen individuals the likelihood of degeneration of the gene pool would be large fairly early in the life of the group.

A further problem is the need for resources. In order to survive, any living creature needs to draw upon what is called a *biomass*. On average, this biomass has to be about ten times its own weight and from this comes the concept of the *pyramid of biomass*.

Imagine a simple system containing foxes, rabbits and grass. This is a pyramid of biomass with foxes at the top and grass at the bottom. The grass is called a *producer* or an *autotroph* because it acquires its food directly from photosynthesis using sunlight. The rabbits exist on what is known as the next *trophic* level and are called *primary consumers* (because they are the first layer above green plants in this particular food chain). In order for the fox community to survive, they need ten times their biomass of rabbits. In their turn the rabbits require ten times their total biomass in

Figure 9.1

Pyramid of biomass for a simple eco-system

grass. This is because for the primary consumers (rabbits) and secondary consumers (foxes), conversion of resources into energy is only around 10% efficient. Without the grass, the rabbits cannot survive and without the grass and the rabbits, the foxes cannot survive.

A group of plesiosaurs in Loch Ness would be at the top of the biomass pyramid for their own *closed system* and in order to survive they would need a substantial mass of smaller creatures beneath them in the pyramid. They may have adapted to an omnivorous diet which would give them a more varied food supply, but even so, a group of creatures their size would stretch the resources of the Loch to its limit, perhaps beyond. But if we argue that the community is very small, we then come up against the problem of lack of genetic diversity. If the Loch Ness monster really is a group of dinosaurs, they would represent a very delicate community. The group could not grow too large because of the lack of resources but could not become so small that in-breeding creates intolerable genetic defects. One of these factors may well have wiped them out long ago.

Those who believe Nessie is a long-lost group of plesiosaurs suggest that the modern creature can live in both sea water and freshwater, so the fact that the Loch gradually changed from a salt-water lake to a freshwater one during the past 7,500 years would not present a serious problem. But the plesiosaur was a marine reptile and it would have spent some time on land; it is surprising there have not been many more witnessed encounters between inquisitive humans actively looking for monsters, and the creatures themselves.

To counter this argument, some believe Nessie is a strange, very large fish, and again this is feasible in evolutionary terms. Sea-water fish could evolve into freshwater creatures as the Loch itself slowly changed its chemical composition once sealed off from the ocean.

So, the Loch Ness monster could be a group of survivalist dinosaurs that have somehow overcome the problems of the biomass pyramid and lack of genetic diversity. They could be marine reptiles that have adapted to freshwater existence or they may be large fish that have performed the same biochemical and anatomical trick. A further alternative is that all sightings of the monster may be accounted for by resorting to logs, tree trunks that have become entangled near the banks of the Loch, or even mirages.

This last suggestion is actually more feasible than the dinosaur enthusiasts would like to accept. The environmental conditions at Loch Ness are surprisingly conducive to mirage formation and these could be produced by objects in the water such as surfacing fish or large branches thrust skyward by freak currents.

Mirages are caused when rays of light are refracted to different extents by layers of air at different temperatures. The eye always interprets light as travelling in a straight line, so if the beam is bent or refracted on its way to the eye, the original image appears to have moved or grown large.

If an object is seen on a cold surface such as the Loch in winter, light rays from the top of the object are bent in one direction by the slightly warmer air and light from the bottom is bent in the opposite direction. This effect is called a *temperature inversion* and although it was only explained during the last century, sailors have reported the illusion it produces since ancient times, calling it 'looming' (hence, 'looming large').

Monster enthusiasts dismiss this explanation as being too limited and point to cases in which bow waves have been seen along with a long-necked creature. But recent meteorological research could also have an explanation for this. The effect of a bow wave or any other disturbance on the surface of the Loch could be produced by what are called *water devils*. These are thought to be smaller cousins of water-spouts, distant relatives of tornadoes, and derive from vortices produced by pressure variations in the water.

These explanations may account for many of the frequent

sightings of Nessie and other lake- and sea beasts from around the world. Fakes and tricks could account for many more, but there is still no definite evidence either way. Loch Ness is vast and impossible to search inch by inch at present. Our best hope lies with future development of super-sensitive sonar equipment or sophisticated thermal tracking and imagining devices that could show what strange creatures may live there, if any.

Almost as famous as Nessie are land-based ECSs. These include the yeti or the Abominable Snowman of the Himalayas and the bigfoot of northern America.

The yeti appears in the legends of the local people of Nepal who worship the beast and will not trespass into its lands. They believe there are three types, the smallest, *yeh-teh* (from which the popular name derives) is about the size of a monkey. The *meh-teh* is taller and heavier, about the size of a small human, and the largest of all, *dzu-teh*, is thought to grow up to nine feet in height.

Modern interest in the yeti began when westerners started to explore the Himalayas and returned home with tales gleaned from the locals and occasionally embroidered with their own fleeting glimpses. Footprints and droppings have been found and many

Figure 9.2

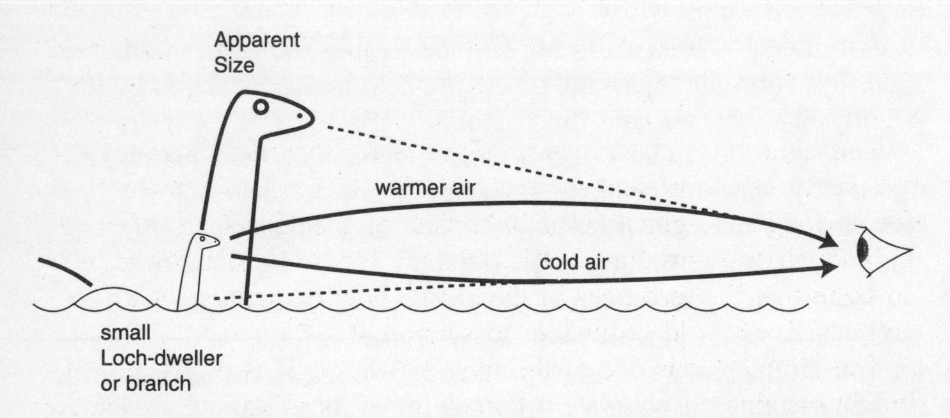

Layers of air at different temperatures cause light rays to be refracted and produce illusions of size.

reputable witnesses, including Lord and Lady Hunt, who took a series of well-publicized photographs of tracks in 1978, have added weight to the claims of believers.

The American version of the yeti is the equally shy, equally massive Bigfoot which is believed to live in wooded areas but has been reported in almost every State. Other similar creatures have been documented in China, in the Urals, where they are known as *yag-mort*, in Siberia where they are given the name *chuchunaa* and in Australia taking the name of *yowie*. Mulder and Scully had some dealings with a land-based ECS in the episode *The Jersey Devil* in which they investigated sightings of a humanoid creature who lived in the woods and ventured into the suburbs under the cover of dark to forage for food.

It is quite possible that there are small communities of people living in the wildest regions who have shunned contact with civilization for generations and there have been documented cases of children found living with animal communities including the famous Wolfboy of France who was found near Avetron in the early 1800s. His discoverer and mentor, a doctor named Jean Marc Itard, tried to introduce the boy to civilization but with only limited success. The wolfboy died at the age of forty still unable to adapt completely to human existence and viewed by many as a freak.

More challenging for science is the idea that there may be groups of creatures living in isolated spots that have evolved in seclusion and about which we have almost no knowledge – true evolutionary cul-de-sacs.

The same rules of biology apply to land-based ECSs as they do to lake monsters or giant sea creatures. For a community to survive for so long it originally had to consist of a sufficient number of individuals to allow for genetic diversity but at the same time, not so large that the resources of the closed environment in which the animals live would be unable to support it.

The Himalayas is one of the most hostile regions on the planet, but it sustains an ecosystem containing a large variety of hardy plants at the bottom of the food chain which support the yak population and other relatively small mammals and birds. A yeti would be at the top of the food chain, but, like the Loch Ness monster, with such a delicate infrastructure, it would live very close to extinction.

In order to maximize resources, the yeti would almost certainly be omnivorous, supplementing its diet of yak and other animals with local vegetation. They may have a thick layer of protective fat and a very thick coat, enabling them to live at high altitudes away from prying humans, only travelling to lower altitudes to hunt. They would almost certainly be solitary, highly territorial creatures that occupy a specific area and meet only to mate (which itself would be a rare event).

The American version of the yeti, the Bigfoot, is at once easier to accept and more problematic for the scientist. Resources, especially food, would be far more plentiful in the woods of the United States and the climate is significantly better. But it is a highly populated part of the world and it would seem unlikely that a creature as large as the Bigfoot of legend would be spotted as infrequently as it has been. In the case of the yeti, it is easy to explain how no bones or remains have ever been located and that footprints are seen only rarely, but this could hardly be applied to a mysterious creature living in Kentucky or California.

With any of these large, humanoid ECSs we also have to address the question of how they could have developed in the first place. Judging by the size of the yeti or the Bigfoot, it would seem they followed an evolutionary path separate from *Homo sapiens* at a far distant point in our evolution. Where the separation could have occurred is almost impossible to speculate. The yeti and the Bigfoot both appear to have more in common with the gorilla than *Homo sapiens*. The gorilla is the largest of the primates and can grow to a height of almost 2 m and weigh 180 kg (400 lbs). They are also very territorial, but usually live in small groups.

The idea that 'humans descended from apes' is a very common misconception. We did no such thing, and despite popular myth stemming from the 1860s, Darwin never said we did. *Homo sapiens* and all other primates have a *common ancestor*. Biologists agree that the evolutionary line that ultimately led to the human form diverged from the ape line during the Tertiary Period. But this is a very broad time span ranging from the end of the dinosaur age (the Jurassic and Cretaceous periods) some 65 million years ago to a time only 2 million years in our past.

Theorists are in some disagreement over when the split between the lineages occurred. Recent paleontological and anatomical evidence from fossil remains suggests that the pongid (ape) and hominid (human) lineages diverged during the Miocene Epoch, which is placed at around 20 million years ago, but evidence based upon comparative immunology experiments points to a divergence as recently as 4 million years ago.

One staggering result from genetic research conducted during the past two decades reveals that the genetic make-up of a human is only very slightly different to that of a gorilla or an orang-utan. In fact, it is now clear there is only a 1% difference between the human genome (the entire genetic composition of a human) and that of a chimpanzee.

All of this would imply that if a type of ape ancestor did diverge from the other branches of the evolutionary tree at some point millions of years ago, its genetic make-up may be very similar to modern primates, including *Homo sapiens*. Of course, we have no way of telling how these creatures might have adapted over thousands of generations. To determine the evolutionary path the creature had followed, biologists would need a sample of blood or some remains from which DNA could be extracted. But if one day such a find is made, it would not surprise most researchers if they discovered the Abominable Snowman is in fact a very close cousin of the Abominable Businessman.

Planet Earth is a very large place. We know more about the far side of the Moon than the bottom of some our oceans and it is certain that humans will live on Mars before they build homes at the bottom of the Marianas Trench, 6 miles (almost 11 km) beneath the waves of the Pacific Ocean. The first modern humans to reach the summit of Mount Everest made the trip under half a century ago and there are many regions of the Earth where no human being has ever walked. It would be complacent in the extreme to rule out the possibility of giant sea creatures, lake monsters and eight-foot tall ape-like beasts. There may be a number of question marks over how small communities of non-interactive creatures could survive time-scales of millions of years, but these do not raise unanswerable questions requiring answers drawn from the occult.

It is revealing that we may only now be starting to encounter these exotic creatures, as we poke and probe into every nook and cranny of the globe. They could all be delusions, man-made

Figure 9.3

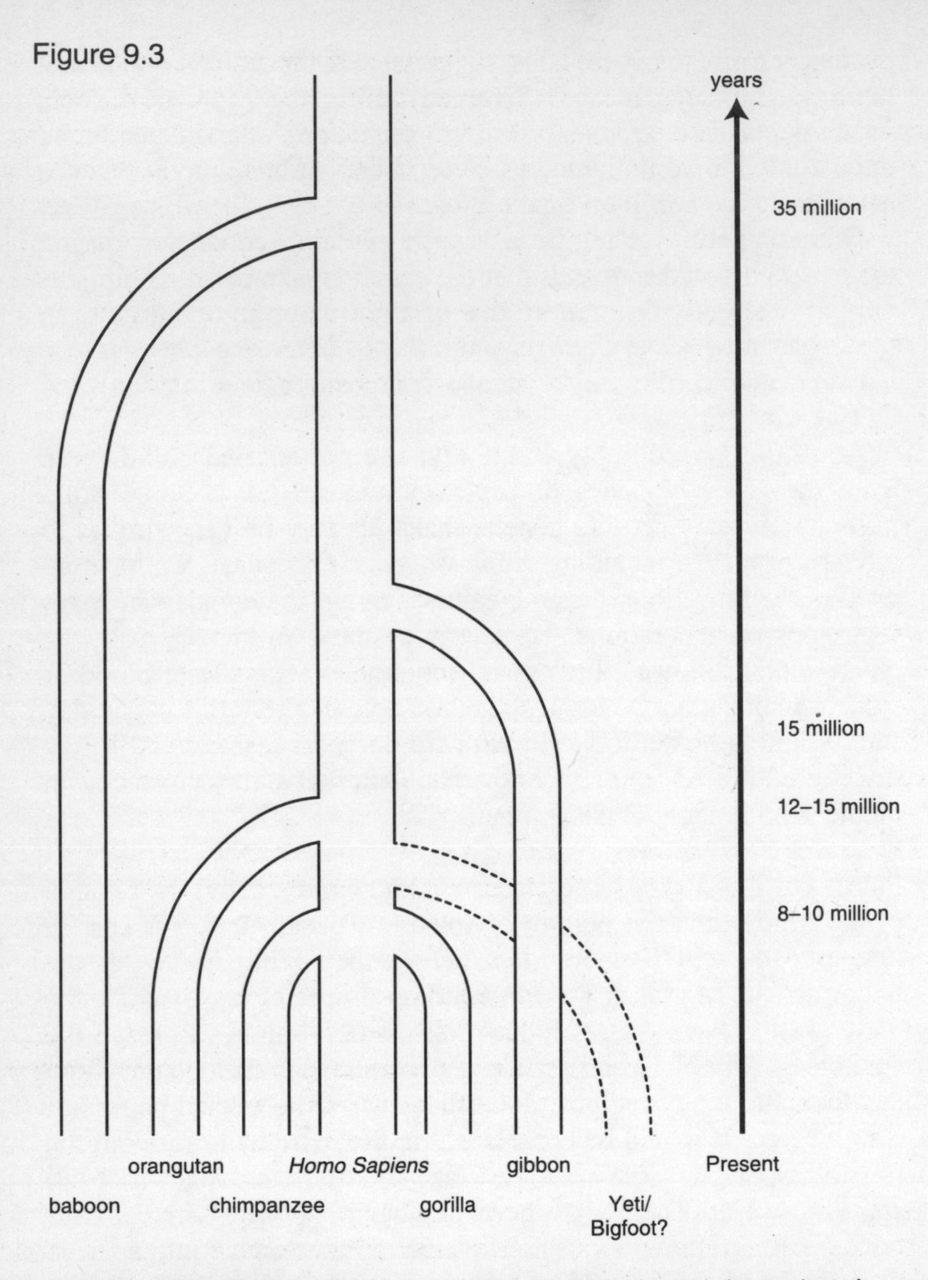

There is considerable disagreement about the time-scales for the evolutionary chart of primates. Fossil records give older figures than immunological studies based upon biochemical research. It is impossible to place even an approximate date at which the evolution of the Yeti diverged, but based upon its apparent physical characteristics, it could have been close to the point at which gorillas diverged from the branch that led eventually to *Homo Sapiens*.

fantasies, but if they are real, we will probably soon know about it. It can only be a matter of time before no place on Earth is secret, no hideaway able to remain beyond our prying eyes. If the yeti, the Bigfoot and the Loch Ness monster do exist, then sadly, their blissful isolation may not last very much longer.

Chapter 10: Time and Again

There was a young lady named Bright,
Whose speed was far faster than light.
She travelled one day,
In a relative way,
And returned on the previous night.

A.H.R Buller[1]

Almost everyone at one time or another has fantasized about time travel. One of the most profoundly frustrating aspects of living in a certain era and limited by our allotted three score years and ten is that we cannot see what will happen in the future; perhaps this is one of the reasons science fiction and fantasy are such popular genres.

Equally, how many times have you wished you could travel back into the past and change something, perhaps some pivotal event in your life? Where would you go in order to correct a problem, unsay something you wished you had never said, undo a wrong perpetrated by or against you? And if you wanted to travel in time simply as entertainment, what point in history would you choose – the Battle of Waterloo or a grandstand seat at Wembley to relive the 1966 World Cup Final?

For the past 300 years, ever since modern science stepped beyond the bounds of guesswork, physicists have believed time travel to be impossible. Yet there have been many documented cases in which people claim to have stepped backwards or forwards in time by seemingly supernatural means. These have been scoffed at, and perhaps for good reason, but since the late 1980s scientists have begun to consider seriously ways in which time travel could actually be possible. Although all the cases of apparent time-slips could be explained away quite easily, it is also conceivable that such things are possible in the natural world and may one day be understood thoroughly or even made practical.

One of the most famous cases of a time-slip occurred in 1901. Two highly respectable elderly spinsters, Charlotte 'Annie' Moberly, who was principal of an Oxford college and Eleanor Jourdain, the headmistress of a girls' school, claimed that during a walk around the grounds of Versailles outside Paris they inadvertently travelled back in time.

It was a hot summer afternoon and the two women were trying to reach the Petit Trianon in the great park of Versailles when they began to see some rather incongruous things. First they spotted a woman dressed in eighteenth-century clothes shaking a white cloth out of a window of a building. Next they passed a couple of officials in greyish-green coats and three-cornered hats. Later they crossed the path of a group of children all dressed in old-fashioned clothes, then encountered a man in a black cloak who had a face badly scarred by smallpox. Finally they came across a woman dressed in fine eighteenth-century dress sitting at an easel and painting. The whole experience lasted about half an hour, and the two women claimed they had reached the Petit Trianon and were given directions out of the park before returning to the world of 1901.

Ten years later, Charlotte Moberly and Eleanor Jourdain published their account pseudonymously under the title *An Adventure*. During the decade between the experience and the appearance of the book they had conducted extensive research into the layout of the grounds at Versailles, the fashions of the period and the history of the gardens. They reached the conclusion they had somehow travelled back 120 years to the time of the French Revolution. They further supported this dubious claim by showing that the landscaping of the grounds had changed since the 1780s and that on their walk they had followed a route that no longer existed. Finally, they believed the 'painting woman' was none other than Marie-Antoinette herself, a conclusion based upon a picture they had discovered in which the French queen was wearing an identical outfit to their phantom woman.

Miss Moberly and Miss Jourdain were respectable establishment figures not likely to deliberately lie or to fabricate such a story even if it meant they would write a best-selling book based upon their experiences, yet there are many problems with their story.

Although they appear to have researched their account thoroughly, they failed to mention some key facts. Firstly, during the time of their visit, a French nobleman, Comte Robert de Montesquiou, spent a great deal of effort and money putting on what he referred to as *tableaux vivants*, (what we would call 'happenings' or interactive art events) set in the grounds of Versailles. These performances involved large groups of the Comte's friends dressed up in the costumes of the eighteenth century acting out scenes from history.

It may have been that Miss Moberly and Miss Jourdain had actually stumbled upon nothing more supernatural than a pageant or role-playing game. They may not have been aware of this, but considering how much research they had done for their book, it seems a strange oversight.

A second factor that casts doubt upon the story is the fact that they saw a woman they believed to be Marie-Antoinette. Despite matching a drawing of the queen with the woman they saw, it seems improbable that if they had travelled back in time they should inevitably encounter so famous a person. But, further discredit came from a letter that appeared in *The Times* in 1965. The author of the letter was a T.G.S Combe who had delivered a lecture on the subject of *An Adventure*. At the end of the talk a member of the audience told him that as a child he was told of an eccentric woman who lived around the turn of the century in the same district near Versailles and on summer afternoons, took to dressing up as Marie Antoinette and sat in the garden of the Petit Trianon.

Most other cases of time-slips are usually less dramatic. Often they involve individuals or sometimes small groups of people thinking they saw a building or a road and when they return to the spot they find the topology of the area is actually quite different to how they remembered it.

One striking case involved four English tourists who stayed in 'a quaint and old-fashioned' guest house in Montélimar in France in 1979. They described the rooms, the other guests and the patron, recounted how they had only been charged less than the equivalent of £2 for the four of them and even took photographs of each other in the hotel rooms. Returning later to see if they could stay there on their return journey, they found a garage on the site of the hotel.

When they had their holiday photographs developed, the mystery deepened. They discovered that those they believed had been taken in the mystery guest-house were missing, but the sequence of the negatives had not been disturbed; it was as if the pictures had never been taken. When asked how it was that the other 'guests' did not seem to be disturbed by their presence in modern dress and stepping out of a modern car, or how they had been able to pay the bill in present-day currency, the two couples were at a loss to explain.

Of course, this case and many others like it could simply be a contrivance, a story deliberately created for publicity or financial reward, but an alternative explanation is that these people had experienced a group hallucination. As we saw in Chapter 8, such things are well documented in cases of hauntings and poltergeist activity. Perhaps the tale covered in *An Adventure* was a collective or group hallucination sparked off by the women seeing the Comte Robert de Montesquiou's *tableaux vivants*. In the case of the two couples on holiday in France, it could be that their genuine hallucination was initiated by too many glasses of the local wine.

An alternative explanation may be due to the play-back system described in Chapter 8. Perhaps events in the past had been imprinted or recorded in the environment and for some reason triggered to replay at the moment these people entered the area.

However, there are some cases in which the idea of a played-back image could not be used as an explanation. This is when witnesses claim to have seen the future. A frequent tale involves people observing long cigar-shaped flying craft or ground vehicles passing by at high speed. Sometimes the occupants are looking out of the windows of their craft and pointing excitedly towards the witness; in other cases, the 'future people' are unaware they are being watched.

Some have suggested that these are visions from a future time projected back into the past, which may occasionally involve a two-way correspondence through which the parties can see one another. Enthusiasts of this idea go on to speculate that such experiences might even account for many unexplainable UFO sightings. This would imply that UFO witnesses do not see alien spacecraft but vehicles from our own far future, perhaps occupied by humans who have evolved into slightly different physical forms.

Much of this is pure speculation. The fact is, we have no idea how a time machine could be constructed and physicists working at the very edge of science are only now beginning to piece together theories that may explain how time travel could be possible. But all of these ideas exist only as mathematical concepts. We are no nearer building a Tardis than we are in reaching the centre of the galaxy in our own time frame.

But the initial step towards building a time machine is understanding the mathematics behind it. And, before we can develop a theory of time travel we have to get to grips with the meaning of time itself.

We all experience the passing of time, but no one seems able to explain what 'time' is. Some even suggest it is nothing more than a construct of our own minds, that we piece together events in a logical, linear order because that is the only way our brains can operate and make sense of the universe.

There is no material evidence to support this concept, and although we have seen already that common sense and cutting-edge physics are often estranged, we all seem to have an in-built awareness of the direction of time, a concept dubbed the *arrow of time*.

Curiously, at the fundamental level, almost all processes in the universe, whether they are interpreted using classical physics or the QM of Schrödinger and Dirac, can almost all be conducted in either temporal direction. This means that if two subatomic particles come together and interact to form two other particles, the reverse process is equally viable; the two product particles could just as well interact to create the starting particles.

Yet, we don't experience this reversibility in the 'real' world, within the macro-cosmos of everyday existence. We don't see shattered glasses reform, light does not leave our eyes and travel to distant objects and the dead do not rise from their graves. So, as we saw with other aspects of QM, it might appear that the principles governing the behaviour of 'simple' systems (those that operate on a quantum level) are in conflict with more complex systems such as those that encroach upon our everyday existence on a macrocosmic level. But even this seems paradoxical because we and every material thing in the universe are made up of fundamental particles. If they behave reversibly in simple scenarios what is it about complex systems that seems to make them act differently?

The answer lies in the difference between something being *impossible* and just very unlikely. Physicists believe that it is not impossible the dead could be made to rise again (ignoring spiritual considerations), or for a broken glass to re-form by chance, it is just that these events require so many improbable steps to juxtapose perfectly (at least compared to the interaction of two sub-atomic particles) that we would almost certainly have to wait for a period longer than the lifetime of the universe to see them happen *naturally*. This means that although they are not *impossible*, they are highly *improbable*.

In order to relate this to the arrow of time, we have to consider one of the most fundamental rules of the universe, a principle called the second law of thermodynamics.

This law lies at the very heart of physics. In his book, *The Nature Of the Physical World*, the physicist Arthur Eddington said that:

> The second law of thermodynamics holds, I think, the supreme position among the laws of Nature. If someone points out to you that your pet theory of the universe is in disagreement with Maxwell's equations – then so much the worse for Maxwell's equations. If it is found to be contradicted by observation – well these experimentalists do bungle things sometimes. But if your theory is found to be against the second law of thermodynamics I can give you no hope; there is nothing for it but to collapse in deepest humiliation.[2]

Unlike some of the exotic aspects of quantum theory and relativity, the second law of thermodynamics is actually a law based entirely upon common sense. Put simply it says that: everything wears out. In more formal terms; the entropy of a *closed system* always increases.

Entropy is the technical term describing the 'level of disorder in a system'. So, by this law, a cup of tea exhibits a higher level of entropy than the individual tea leaves, water or milk because they have been mixed together. It would take more energy to separate them out again than was used to mix them in the first place.*

* Our universe is a closed system, so entropy will always rise in the universe.

Returning to the example of the broken glass. If we tried to bring together the pieces, like running a film backwards to recreate the glass perfectly, we would need to lower the entropy of the system. This is possible (in fact, living creatures spend most of their lives attempting to produce a local lowering of entropy), but it requires energy and for this to happen naturally by chance is incredibly unlikely.

In a similar way, a garden left to overgrow will gradually increase its entropy level quite naturally. In order to restore the tangled weeds and vines to their former order, work would have to be done, or energy expended. It is extremely unlikely this will happen naturally without the interference of intelligence (and muscles).

Because the natural processes of the universe are all ones in which disorder or entropy is seen to be increasing, it gives us an indicator, a way to view the progress of the universe, or in other words, the direction in which time flows.*

So, if there is a definite direction to time, can intelligent beings or even information move in a non-linear way from one time frame to another? Would it ever be possible to build a machine like the Tardis?

Currently, physicists are giving serious consideration to two possible mechanisms via which a genuine time link could be produced. The first of these is to employ the ever-useful wormhole.

We saw in Chapter 6 that using a wormhole to transport information may be within the bounds of acceptable physics utilizing Einstein's theory of relativity. In that discussion I used them to explain how precognition could be a genuine natural process, but this system might also be used to create a two-way time-travel system.

In the example, the observer in 'our time' viewed the other end of the wormhole as the future and they could witness events that had not yet happened, but of course, to those at the other end of the wormhole, we exist in their past. So, a wormhole could produce a

* We might even say that although it would seem that intelligent life is constantly attempting to decrease entropy locally, life itself could be the very reason the universe will one day end. This is because every time a process occurs there is a loss of *useful* energy from the system (energy that cannot be harnessed to do work). As the physicist Barrow Chapman has said: 'It may be that the purpose of life is simply to facilitate the heat death of the universe.'[3]

time loop – travel one way through it and you emerge in the past, take the opposite direction and you arrive in the future.

Although the evidence for the existence of black holes is considered very strong, wormholes could prove to be nothing more than imaginary entities. It would be a far less interesting universe if wormholes and time-loops did not exist, but sadly, the laws of physics are not malleable to human desires. Conversely, if in the future black holes are shown to be common, then it is very likely that wormholes could exist and if we advance sufficiently to first transport information and then objects through these temporal and spatial highways, time travel could become a real possibility. Perhaps others are already doing it.

Stephen Hawking and others have speculated that mini-wormholes exist naturally in our universe and it may be these that are responsible for the accounts of time-slips. It could be that the information content of an incident at the antiquated hotel in Montélimar was somehow transported to a future time quite naturally and was observed by the four British tourists who claimed they had visited the place. We saw in Chapter 6 that some enthusiasts have suggested that particularly traumatic events may somehow trigger wormholes or the energy from such happenings could utilize naturally occurring wormhole links between different times. Perhaps these amazingly handy devices are more common than we think. It is even possible that our universe is interlaced with a vast network of interlinking wormholes just waiting to be employed by a sufficiently advanced technology.

There is another possibility which involves black holes but does not require the existence of wormholes. This is the idea of 'skimming' the intense gravity well of a black hole.

Again this idea is based upon the equations of Einstein's theory of relativity. The first mathematician to seriously speculate upon the possibility of time travel using the equations of relativity theory was a friend and colleague of Einstein's at the Princeton Institute for Advanced Study, Kurt Gödel in 1949. Fourteen years later, a New Zealander, Dr Roy Kerr published a paper speculating upon the idea of a time machine using the theory of relativity as applied to black holes. At the time, black holes were still unchristened (that honour fell to John Wheeler in 1967) but Kerr knew such objects were feasible in principle and employed the fact that time is affected by velocity and gravitational fields to demonstrate a theory of time travel. By an amusing coincidence his paper was published on the eve of the first episode of *Dr Who* in November 1963.

Kerr's theory suggested that if a time machine was fired at a black hole and made to skim the edge of the gravitational well without being sucked in, time would travel far slower for the occupants of the machine. Meanwhile, the events in the world outside would be whizzing by. If the machine then travelled back to a point beyond the black hole they would find themselves in the future.

A decade later, the concept of using black holes as time machines was extended by Frank Tipler from the University of Maryland, and in 1974, he detailed his ideas in a paper published in the highly respectable journal *Physical Review*.[4]

Tipler took things much further than Roy Kerr. In his scheme, a very advanced civilization could produce a special type of black hole called a *naked singularity*. To make this, the singularity (found at the heart of a black hole) would have to be rotating. The effect of the rotation is to twist space-time so much in the region near the singularity that time itself becomes another dimension of space through which a carefully piloted craft could be manoeuvred.

Tipler then went on to detail the design spec for the artificial naked singularity. According to his calculations you would need a cylinder 100 km long and about 10 km across made of super-dense material, similar to that found in a neutron star where all the electrons of the atoms of the substance had been fused with the protons in the nucleus. Finally the object would have to spin precisely twice every millisecond.

For all the ingenuity of this scheme, on its own it might be considered nothing more than a fantasy, but amazingly, there are naturally occurring objects in the universe that almost fit the bill.

In Chapter 2, I described the story of how the British astronomer Jocelyn Bell discovered the first pulsar in 1967 by tracing a regular radio signal which was believed initially to be a message from an alien civilization. Although this signal turned out to be a natural pulse, the consequences of its discovery may be every bit as exciting as if it had been a beacon placed there by an alien race. This is because special objects called *millisecond pulsars* have since been discovered that are so close to being Nature's time machines they may only need slight adjustment by an advance technology to be usable. Millisecond pulsars are made of material with almost the right density and they spin once every 1.5 milliseconds (one third the speed needed for Tipler's design).

Even though we may be thousands of years away from having the technology to utilize such objects, the discovery of millisecond pulsars combined with the innovative ideas of Frank Tipler and others is now generating great excitement within the physics community. But there remains another delicate matter – the problem of temporal paradoxes.

The concept that time travel can create agonisingly complex paradoxes has been known for centuries and long before any serious thought had gone into how a time machine could be designed, it was believed that these paradox problems could actually be so severe they alone would prohibit practical time travel.

H.G. Wells set the tone with his classic novel *The Time Machine*, published just over 100 years ago in 1895.[5] Unless Wells had a time machine himself, he would have known nothing of relativity because the creator of the theory, the sixteen-year-old Albert Einstein, had just squeezed his way into a technical college in Zürich at the time and was struggling with elementary maths. Not surprisingly, the author offered little by way of explanation for his time travel system, but he was careful not to send his hero into the past almost certainly because of the problematic paradoxes such a journey could entail.

Happily, the problem may not be as bad as Wells and others feared. By utilizing some intriguing aspects of quantum theory, physicists are now concluding that temporal paradoxes could be completely fictitious.

In his short story, 'All You Zombies', written in 1959, Robert Heinlein offered what must be one of the most confusing examples of a time travel paradox ever imagined.

The story centres around a character called 'Jane' who is mysteriously abandoned at an orphanage in 1945. The child grows up with no idea who her parents are, but in 1963, at the age of eighteen, she falls in love with a drifter who visits the orphanage. For a while things go well, but then the drifter leaves her and Jane finds she is pregnant. The delivery of the child is difficult and she has to undergo a Caesarean. Then, during the operation, surgeons discover Jane has both sets of sex organs and in order to save her life they have to convert 'her' to a 'him'.

Subsequently the baby is mysteriously snatched from the hospital, Jane drops out of society and finally ends up a vagrant. Seven years later, in 1970, he stumbles into a bar and becomes friendly with the bartender who offers Jane a chance to avenge the drifter who had ruined his life on the condition that he joins the 'time travellers corps'. The pair then go back in time to 1963, the vagrant 'Jane' seduces the eighteen-year-old female Jane at the orphanage, making her pregnant before disappearing. The bartender then travels forward in time nine months, snatches the baby from the hospital and deposits it at the orphanage in 1945 before dropping off Jane in 1985 where he joins the time travellers corps which has been created after the recent invention of time travel.

Jane, the time traveller, distinguishes himself in the corps, and eventually becomes a highly successful bartender, opens his own place in 1970 to persuade a young vagrant to join the time travellers corps.

So, in this tale, Jane is her own mother, father and daughter. She is also the drifter and the bartender. But who are Jane's grandparents? She seems to be a creature of time, self-created and totally independent of the universe, in other words, a paradox.

There are other simpler examples of this twisting of events. Imagine a time traveller journeying 100 years into their past to the studio of a struggling artist. There, they tell the artist that in the future, he is world famous, recognised for a distinctive style very different to the one he is currently using and then proceeds to show him a catalogue of his future work. Distracting the visitor, the artist photocopies the artwork and the time traveller returns to the future. The artist then starts to copy the paintings he has photocopied.

The disturbing thing about this paradox is that it seems to offer a free lunch and taken on face value, it breaks the laws of physics. Which came first, the paintings or the artist's fame? It also seems to cancel out the principle of free will. If beings from the future are able to manipulate the past and change our lives, where is the element of self-determination?

Fortunately, there is a solution to this set of possible paradoxes utilizing a concept physicists call the many-universes interpretation.

I touched on this theory in Chapter 6 when discussing various interpretations of the Schrödinger's cat experiment, but the implication of the idea may have even stronger relevance to the subject of time travel.

The simplest interpretation of the many-universes theory is that whenever any fundamental event occurs, the future splits into two possible outcomes or separate universes. It is easy to understand this when we refer it to our own lives. Suppose we have an important job interview. Perhaps along one route we make the interview, get the job and eventually become the chairman of the company. In the other, we miss the train, fail to make the interview and lose the chance of a perfect job.

But, this example is one on a macro-cosmic scale. According to the many-universes interpretation, every time any sub-atomic change occurs anywhere in our space-time continuum, the path splits creating two different universes. These universes may be so similar that any difference may be completely imperceptible to us. Perhaps the only variation is the position of one electron situated the other side of the universe. Even so, they will be different. And it is because of this that the troublesome time travel paradoxes could be written out of the equation.

To visualize this consider another example. If in Universe A, 'our' universe, we go back in time and persuade our grandfather not to go on a crucial date with our future grandmother, a paradox will be avoided because, at the instant we arrive in the past, two possible futures or universes are created simultaneously, Universe A and Universe B. In Universe A, our grandfather goes out to dinner and starts a lifelong relationship completely unaware of our arrival. This leads to a future in which we are born and become a time traveller who returns to the past. In Universe B, the grandfather misses the date and we are never born. But, because we have come from Universe A, we do not suddenly cease to exist and there is no paradox.

Stephen Hawking used to say that if time travel was really possible we would be visited by time tourists; but as we are clearly not, it is impossible. This philosophy is wrong for at least three

Figure 10.1

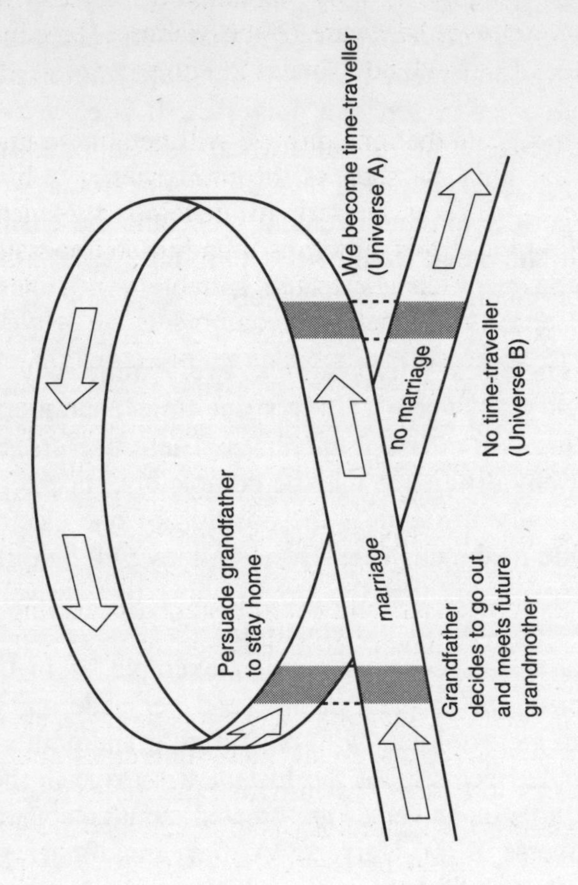

We become time-traveller
(Universe A)
No time-traveller
(Universe B)
no marriage
marriage
Persuade grandfather
to stay home
Grandfather
decides to go out
and meets future
grandmother

reasons. Firstly, time travellers would almost certainly be sophisticated enough to cover their tracks. Secondly, our space-time constitutes only a vanishingly small part of the entire life and volume of the universe, so it is highly probable time travellers have not yet visited this time or place. Thirdly, if the many-universes interpretation is correct, only versions of ourselves in certain universes would ever be aware of the visitors. Hawking has since changed his mind and now believes time travel is theoretically possible.

It seems probable that one day we will be able to utilize natural aspects of the universe such as the properties of a black hole or exotic objects such as pulsars to develop a device to travel backwards in time If this system is found to be impossible then we might still have the chance to use suitable wormholes. Either of these will require dramatic developments in physics and in particular a successful combination of quantum theory and relativity which remains the Holy Grail of modern physics.

Whether or not fluke conditions sometimes arise on Earth enabling people to wander into the past or the future by accident is still open to conjecture. Given the tremendous forces at work within pulsar time machines or wormhole temporal highways, this would seem unlikely, but then there are almost certainly other exotic objects like pulsars and neutron stars just waiting for curious scientists to discover them. Perhaps some of these will be located nearer home.

Finally, the ideas of quantum mechanics seem capable of solving the problem of paradoxes. If the many-universes interpretation is proven one day to be the way in which the multiverses of creation really behave, and the technology is in place, then the last barrier to time travel really could be overcome. Perhaps it is premature to book a ticket on the Temporal Express, but we can dream.

References

1. Professor Arthur Buller, *Punch*, 19 December 1923.
2. As quoted in John Gribbin, *Companion to the Cosmos*, Weidenfeld & Nicolson, 1996.
3. Barry Chapman, *Reverse Time Travel*, Cassell, 1995, p. 107.

4. Frank Tipler, 'Rotating Cylinders and the Possibility of Global Causality Violation,' *Physical Review 9D*, 1974, pp. 2203–6.
5. H.G. Wells, *The Time Machine: An Invention*, Heinemann, 1895.

Chapter 11: Into The Light

> They were seated in the boat. Nick in the stern, his father rowing. The sun was coming up over the hills. A bass jumped, making a circle in the water. Nick trailed his hand in the water. It felt warm in the sharp chill of the morning.
>
> In the early morning on the lake sitting in the stern of the boat with his father rowing, he felt that he would never die.
>
> Ernest Hemingway[1]

We have countless euphemisms to deal with it – passing away, kicking the bucket – but, whatever we call it, one day we will each face the Big D. or as the geneticist Steve Jones has said recently while discussing the odds of dying from various causes: 'Not every gambler will win the Lottery, but all of them will die.'[2]

Morbid? Yes, and that is why we humans have tried since members of our race first became sentient to figure out ways to convince ourselves that death is not the end, that there is an Afterlife, be it in heaven, hell, or right back here again. But, what evidence is there? Is it all wishful thinking as materialists would claim, or is there after all something special about us as individuals, something immortal?

Until the 1970s there was little attempt to make a psychological study of the process of dying; naturally, those who had died could tell us nothing. But gradually, stories were documented which involved individuals travelling to the brink of death, even being declared dead, only to return to life. Some of these people were able to recall what they had felt and the term Near Death Experience or NDE was coined to describe the process.

The first person to popularize the idea of the NDE was a doctor working in Georgia called Raymond Moody who wrote a book on the subject called *Life After Life*[3] which has since sold over 3 million copies. Although Moody made no attempt to analyse or judge the experiences he described, through a collection of rea cases, he introduced into the language many notions synonymous with NDE – seeing one's own body from above, travelling along a dark tunnel to the light beyond and meeting a spiritual guide. The first serious analysis of these experiences and an initial attempt to quantify claims came with the work of another American doctor, Kenneth Ring, at the University of Connecticut.

In 1980, Ring categorized five separate stages of the Near Death Experience. These he labelled: peace, body separation, entering the darkness (or tunnel), seeing the light and entering the light. The majority of individuals who have reported an NDE only recall the initial stages and the full five stages have been experienced by fewer than 10%.[4] Yet strikingly, most analysts believe there is a certain uniformity in many of the descriptions and a typical full experience might consist of the following details:

> I was suddenly aware of a calmness, a still, peaceful feeling. Then, all at once, I felt myself floating above my body and looking down at the proceedings, totally detached, as though I was watching a movie. I could see myself from above, I could see the tubes and pipes and doctors around me, the flatline read-out on the screen.
>
> Gradually I became aware of a darkening around me, as though the scene was narrowing and then I could see a tunnel with a bright light at the end of it. I felt compelled to move towards the light and felt myself accelerating that way. As I travelled I could see scenes from my past flashing before me, both good and bad moments. Then, as I approached the brightness, I could see someone; they were surrounded by light. Suddenly, I recognized the figure as my father. He had died ten years earlier, but there he was, just as I remembered him. He was asking me if I was ready for this journey. Did I want to go on with him or go back? Then, a moment later, before I even had time to answer, I felt myself being dragged, sucked back into my body. I snapped awake and the agony returned in a great rush. I felt terribly angry and then gradually calmed down.

To many, descriptions like this are clear evidence that we live on in some form after physical death. But sadly, upon close scrutiny, such claims may be quite unfounded. Let us take each stage of this amalgamated description and consider the possibilities.

The sense of relaxation, or impression of peace, is perhaps the easiest aspect to explain using accepted medical knowledge and almost certainly comes about because the body is being flooded with 'calming chemicals' such as endorphins. These are naturally-occurring agents produced by the brain that serve to relieve pain during times of extreme physical stress. Fitness enthusiasts make use of them in the gym – it is the endorphin rush that keeps them going, and by controlling it athletes can push themselves through the pain barrier during training. It is no surprise these chemicals are released, it is another example of how the body switches into survival mode during physically demanding moments.

The next stage of an NDE involves the patient seeing themselves and their surroundings. This is almost without exception from a bird's-eye perspective, looking down upon the proceedings with feelings of detachment. This stage of the NDE has been given the name Out of Body Experience or OBE.

Claims of OBEs are not restricted to traumatic situations such as a Near Death Experience. There are many people who claim they have become dislocated from their bodies and have been able to wander around in an astral world. In many cases this behaviour seems to be indistinguishable from dreaming. One astral traveller, Robert Monroe, has described vivid tales of his trips to the planets of the solar system and beyond to the distant stars. He claims to be able to travel across intergalactic space outside of time. Unhindered by any of the tiresome restrictions of relativity, he travels instantly to any location and supports his stories by suggesting that he has seen 'markings' on planets before NASA spotted them using satellites.[5]

Naturally, there is little to back up these far-fetched stories, but they do make good material for best-selling accounts. More interesting, but a little less dramatic are reports from individuals who claim to have had OBEs whilst in a traumatized situation and as part of a Near Death Experience. These are intriguing because they are part of a process and not apocryphal recollections or one-offs that offer little chance of rational explanation.

Much has been made of the OBE aspect of the Near Death Experience. This is perhaps because it is the only stage that can be related to the living world and other individuals. Often victims report snatches of conversation heard from close by or can quote what doctors have said as they were making attempts to resuscitate them. They are also able to describe bits of machinery surrounding them and the sounds they made. Very occasionally a patient will claim to have seen something that has not occurred in their immediate vicinity or to report an incident in an adjoining room and it is these apparently supernatural abilities that have most excited believers in life after death.

It is actually not surprising that people suffering trauma can recall some of the things they have seen and heard around them under these conditions. It has been found that people are quite capable of recalling things that have been said nearby while unconscious or asleep and events that occur in the 'real world' close by often appear in the individual's dreams.[6]

They do not need to be astrally projecting to achieve this. Probably the most common example is the behaviour of new parents towards their baby. The slightest sound can wake them from deep sleep if it is associated with the child, but extraneous, unconnected auditory signals are ignored. Again, this is a survival mechanism dating from primitive times, a deeply rooted instinctive reflex.

There have also been extensive studies of anaesthetized patients who have been found to register external stimuli although it is often found that many do not recall the details afterwards.[7]

But all of these rationalist arguments could be overridden if there was proof that an individual in the midst of a NDF was able to perceive information they could not possibly have gained through normal physical means. Unfortunately for the enthusiasts, when studied closely, any 'evidence' melts away. The psychologist Susan Blackmore has made a thorough study of OBE claims and has analysed several hundred cases, yet not one stands up to close scrutiny.

One of the best examples comes from a cardiac patient, a woman named Maria who recovered from a heart attack and later described how she had floated over her body and travelled to another room in the hospital where she saw a tennis shoe left on the window ledge. Later, a nurse went to look for the shoe and found it exactly as described. However, Maria is now dead, the shoe was not been seen by anyone else and the only support for the claim comes from the singular testimony of the nurse.

Another story describes how a woman claimed she had left her body and saw her son, Graham, arguing with a nurse before he was told to leave the room. The woman then travelled astrally above her son as he walked along a corridor. She saw him enter a room where his partner, Chris was waiting. Chris asked Graham what had happened and he replied that they would not let him see his mother. He then kicked a chair before sitting down and lighting a cigarette.

The problem with this is that every aspect of the story is predictable. Despite being in a state of trauma, the woman's brain could have registered her son arguing with the nurse. She then imagined the logical series of steps that would follow such an incident. She would have known who was most likely to be at the hospital, she would have been able to predict her son's likely reactions (kicking the chair) and she may have further assumed he would light a cigarette. There is no proof here, nothing to pin down or to demonstrate anything that could not be explained by quite natural means.

Yet there are some intriguing aspects coming out of experiments conducted by doctors who have treated patients who believe they have had a NDE. One interesting piece of research comes from cardiologist, Dr Michael Sabom. He has conducted experiments to see just how much can be remembered of their environment by those who claim to have recovered from an OBE. He asked these individuals what they could recall of their own anatomy and the function of various pieces of surgical apparatus in the operating theatre. To his surprise, he discovered that those who claimed to have had an OBE were considerably more knowledgeable than a test group who had undergone the same surgery without reporting an OBE.

On face value, these could be viewed as circumstantial evidence, but so little is known of the brain state during what patients believe to be an OBE that we cannot jump to conclusions. It could merely be that patients who do have what they perceive as a mystical experience actually experience a mental state that makes their short-term memory far more receptive. In this way, they could absorb more information about their environment than people in a less traumatized state.

One of the most intriguing elements of reported cases of NDE is that in the overwhelming majority of cases, the patient 'sees' themselves from above. The almost universal character of this aspect of the near-death experience has led enthusiasts to use it as evidence that an astral body really does detach itself from the physical. But in fact this universality is quite suspicious. Why, if the spirit or the astral body is leaving the physical body, should they only see their physical body from above?

Psychologists believe that this 'floating above the body' perspective supports the conclusion that all OBEs are self-induced hallucinations or visualizations originating from a traumatized brain. The reason for this is that as living, breathing individuals, we build models of the world as we see it through our eyes at the time we experience it. So the world is always visualized at eye-level. But, in our memories we almost invariably visualize a scene either from a bird's-eye view or from the viewpoint of a third party. We very rarely imagine or remember a scene in which we played a part from our perspective at the time, despite the fact that almost all the audio-visual information gathered by our brains comes from this angle.

When our brain is dying and the audio-visual impulses are becoming confused or are largely shut down, we turn to our memories and dreams to construct a model of the world, and these images are almost always seen 'from above'. This is the viewpoint of the dreamer, those in a trance-like or soporific state and those suffering from severe brain trauma surviving at the limits of consciousness.

Susan Blackmore at the University of Bristol and Harvey Irwin from the University of New South Wales, Australia have conducted experiments on sets of people who claim to have had OBEs and have found that in many cases these individuals are particularly good at switching viewpoints. Most importantly, they report their dreams to be largely perceived from a bird's-eye view and can more easily imagine a waking scene from this angle.[8]

After this sensation of floating above the body, most of those who remain in a near-death state report that their vision becomes restricted and their entire environment appears to be taken over by a tunnel at the end of which they can see a bright light.

Believers in life after death perceive this to be a 'passage to the afterlife', at the end of which is a heavenly light, a place where we will meet those most important to us and decide either to travel on to an existence beyond our physical lives or return to recover here on Earth. But again, work conducted in recent years by neurophysiologists and psychologists is casting doubt on this very attractive idea.

One suggestion for the origin of this tunnel came from Carl Sagan who proposed that the experience was a 'birth memory', but this idea has fallen out of favour recently.[9] Psychologists point out that birth memory is very rarely a fully formed notion and that being born is actually nothing like going through a tunnel. It was then proposed that the tunnel was a *visualization* of birth – an image the brain created in order to picture what birth was really like. The idea was finally disposed of with a survey in which 254 people were asked if they had ever had a Near Death Experience. Out of these 254, 36 of them had been born by Caesarean section, but almost equal numbers responded with a conviction that they had indeed endured a NDE.[10]

A more likely explanation is that the brain is suffering what is called *cerebral anoxia* or oxygen deprivation. It is this purely physical phenomenon, psychologists now believe, that is responsible for the tunnel and the light. It is also thought to lead to the frequently reported encounters with long-dead relatives or even a godlike figure.

As discussed in Chapter 3, the brain operates by electrical impulses passing along neurones. These impulses cross gaps called synaptic gaps between the ends of neurones. The messages that pass across these gaps have to be carefully controlled. Too many impulses too rapidly, and the signals cannot be processed properly; too few, too slowly and brain function is impaired. This means the inhibition of signals through the brain is every bit as important as excitation.

It was found in the 1950s that hallucinogenic drugs like LSD and mescaline operated by quashing the inhibitors in the brain so that signals passed across the synapses with increased speed and intensity. LSD suppresses the action of the *Raphe* cells which regulate activity in the visual cortex, a part of the cerebral cortex or the thought-processing centre of the brain.

Experiments conducted on rats have shown that when a brain is deprived of oxygen the inhibitor signals are shut down before the excitor signals, so the nerve cells pass on their impulses more rapidly and efficiently. This will have two dramatic effects. Firstly, just as happens with LSD, reception of impulses will be altered in the visual cortex. Secondly, cerebral anoxia produces hallucinations.

The visualizing of a tunnel had been well documented long before the NDE had been popularized and was known to be part of an hallucinatory experience created with drugs such as psilocybin (magic mushrooms) and LSD. In the 1930s, a psychologist at the University of Chicago called Heinrich Kluver noted that there were four basic images seen during drug-induced hallucination. These are the lattice, the spiral, the cobweb and the tunnel, but it has only been during recent years that the reason for this limited collection has been explained.

Neurobiologist Jack Cowan worked on this problem during the early 1980s and came up with a link between signals registered by the retina and the images visualized in the brain.[11] He concluded that a traumatic disturbance in the brain such as oxygen deprivation causing loss of inhibition signals would create what he called stripes of activity in the visual cortex. When these stripes were translated into a perceived image they appeared as concentric circles.

More important still is the fact that when we register information on the retina, processing the visual data from the centre of the visual field requires the most neurones. If the brain has been starved of oxygen and the inhibitor signals are suppressed, this will affect all neurones equally, but because the centre of the visual field requires more neurones this will suffer the most dramatic disruption – hence the bright light at the end of the tunnel.

Fortunately, this theory can be verified using computers. If a processor is programmed with the information needed to recreate a simple model of the link between the retina and the visual cortex and then a simulation of increased excitation is added, a tunnel-like image is produced on the screen. This image is bright at the centre and darkens at the edges. If the level of activity is increased to replicate the effect of increasing the cerebral anoxia, the 'tunnel' appears to come closer which creates the illusion of moving along it towards the bright light at the end.

Some very intense experiences have been reported by those brought back from the very edge of death. They sometimes report that as they enter the light (the fifth and final stage of any reported

NDE), the tunnel image appears to break up and occasionally transforms into fields, trees and other distorted images. Amazingly, this is exactly what happens in the computer simulation. If the stimulus is increased to represent further oxygen loss and greatly increased inhibition of signals, the tunnel transforms into a set of spirals and complex rings.

But what of the overpowering impression that as the patient emerges from the tunnel and enters the light, they often encounter lost loved ones or even in some cases, the Creator?

It seems likely that these images are also created by cerebral anoxia because, as the normal brain functions become confused or overloaded, we are forced to call up dreams and memories to deal with the situation. As Susan Blackmore describes it: 'The system simply takes the most stable model of the world it has at any time and calls that "reality". In normal life there is one "model of reality" that is overwhelmingly stable, coherent and complex. It is the one built up from sensory input. It is the model of "me, here, now." I am suggesting that this seems real only because it is the best model that the system has at the time.'[12]

People severely injured or suffering a heart attack or other such medical emergency would naturally cling to the most secure of their fantasies or dreams. The most likely image to appear in these hallucinatory states would be lost parents, partners, or in some cases, an image of their God.

One of the most striking aspects of NDE and one which stimulated the hopes and imaginations of believers very early on is the claim that the experience is uniform, that the images are universal and cut across all religions and cultures. But according to some critics, this is not entirely true.

Firstly, they point to the fact that many people do not report their experiences from fear of ridicule; this lowers the sample size considerably. Even then NDE seems to be a very rare phenomenon. In his book, *Pseudoscience and the Paranormal*, the sceptic Terence Hines claims that there are a number of cases in which people report very unpleasant Near Death Experiences analogous to the occasional 'bad trip' recalled by users of hallucinogenic drugs.

This may be caused by a variety of factors, including the brain chemistry of the patient at the time of the NDE and obviously the reason for them being there in the first place. It could also be influenced by the patient's mood, life situation or the medication they are receiving as the doctors try to bring them back from the dead.

Most psychologists agree that the reason why the majority of NDEs share a set of common elements, irrespective of class, race, sex or religion is simply because similar brains create similar images. This would imply that many aspects of personality – a person's tastes, interests, their job or their preoccupations – are actually 'non-core' or peripheral activities that have little effect on the way their brain behaves during shut-down. It would appear that the deeper drives, strong emotions – their loves and hates – play a far greater role in the creation of hallucinatory images both drug-induced and trauma-associated.

No one who has endured a genuine Near Death Experience has been left unaffected by it. Many people believe it to have been a pivotal event in their lives. Some have been changed beyond recognition by the experience; for most it has radically altered their perception of life and their place in the scheme of things. Almost everyone (excluding those unfortunate enough to have had a bad NDE) has come away from it no longer sharing the fear of death so natural to most others. Some have even claimed to be looking forward to dying.

Science can show that the claims of those who have lived through a NDE are not based upon the supernatural. These people are not treated to a glimpse of an Afterlife, but they are nevertheless very lucky for two other reasons. Firstly, they survived, but also, they experienced a very rare phenomenon and one most people could only hope to experience just before they die. There is no doubt that NDE is a genuine and fascinating phenomenon. Those who claim to have seen a tunnel, even those who believe they met God for a moment, are not lying; they believe genuinely that they experienced the real thing. What they did perceive was a representation of powerful, personal images associated with life and death, all within their own cerebral cortex.

Perhaps the Ancients knew about the NDE. Perhaps the legend of heaven and hell has grown out of the 'good trip' and the 'bad', each stimulated in part by our deepest drives, desires and fantasies.

Maybe this is the real reason why believers have always declared that the good of heart go to heaven and the wicked go to hell.

The NDE is a reassuring discovery. It is comforting to think that we can at least hope to have this naturally induced solace when our brains close down forever. Many die instantly, by a wide variety of means, but perhaps the greater number of all the humans who have ever lived simply slipped away from physical existence. And if comforting chemicals and cushioning brain processes create dreams and hallucinations, it may be thought of as a neat biological trick to aid our passing.

For many, life may be cruel, but perhaps it's not entirely malicious.

References

1. Ernest Hemingway, 'Indian Camp', published in *In Our Time*, Cape, 1926.
2. Steve Jones, 'View From the Lab', *Daily Telegraph*, 5 June 1996.
3. Dr Raymond Moody, *Life After Life*, Mockingbird Books, 1975.
4. Kenneth Ring, *Life At Death: A Scientific Investigation of the Near-Death Experience*, New York, Conard, McCann and Geoghegan, 1980.
5. Robert Monroe, *Far Journeys*, Souvenir, 1986.
6. *The Mind in Sleep: Psychology and Psychophysiology*, Edited by A. Arkin, J. Antrobus and S. Ellman. Hillsdale, NJ, Lawrence Erlbaum Associates, 1978.
7. K. Millar and N. Watkinson, 'Recognition of Words Presented During General Anaesthesia', *Ergonomics, 36* (1983), pp. 585–594.
8. Susan Blackmore, *Journal of Mental Imagery*, Vol 11, 1987, p. 53.
9. Carl Sagan, *Broca's Brain*, Random House, 1979, p. 143.
10. Susan Blackmore, 'Birth and the OBE: An Unhelpful Analogy', *Journal of the American Society for Psychical Research, 77* (1983), pp. 229–38.
11. J.D. Cowan, 'Spontaneous Symmetry Breaking in Large-Scale

Nervous Activity', *International Journal of Quantum Chemistry, 22* (1982), pp. 1059–82.

12. Susan Blackmore, 'Visions from the Dying Brain', *New Scientist*, 5 May 1988, pp. 43–45.

Chapter 12: The Healing Touch

> There was a faith-healer of Deal
> Who said, 'Although pain isn't real.
> If I sit on this pin,
> And it punctures my skin,
> I dislike what I fancy I feel.'
>
> Anon[1]

'Rise up my child, you're cured,' the evangelist shouts, and the huge audience cries out in harmony, 'Hallelujah'. Then on the other side of the stage a man crumpled into his wheelchair suddenly shouts out 'Praise the Lord' and eases himself from the chair he has not left for over a year, takes a few tentative steps towards the preacher and the crowd roars.

At first glance such dramatic shows seem very impressive; the evangelical healer would claim that God had given him or her the power to repair damaged limbs to eradicate cancer and to restore sight to the blind. But how truthful are these claims and is there any scientific substance to any form of faith or spiritual healing?

Faith healing is one of the oldest mystical arts and has a place too in orthodox religions. Jesus was supposed to have returned Lazarus to the living and, according to the Bible, the sick could be instantly cured by simply touching the hem of Christ's robe. Throughout the Dark Ages and into the modern era, countless quack doctors and miracle workers have claimed to possess the healing touch whether it is a form of internal power or because they act merely as conduits for their all-seeing, all-knowing, all-caring gods. The healer is another example of an image that has become so ingrained that it is an archetype. Perhaps this is one of the reasons why even discredited evangelists maintain such huge devoted followings.

Healers, whether they enwrap their practices in religion or use it in its purest form, fall into several categories. The two most important types are those who claim they channel energy from beyond themselves to the patient and those who believe the energy they use comes from within them.

The typical evangelical healer is of the first type and engenders the belief in his followers that his powers come from God. And in the majority of cases followers believe the healer because they desperately need to. Naturally, the preacher/healer (who is actually only interested in making money from the hapless members of his audience) emphasizes this need and exploits it to the very limit.

Massive shows such as those put on by the evangelist Morris Cerullo at auditoria the size of Earls Court or Madison Square Garden are such gargantuan events they generate the near-hysteria usually associated with football matches or pop concerts. In this environment of religious fervour and deep-rooted need to believe, it is easy to see how people become even more open to suggestion, more readily manipulated.

Shows like Cerullo's are carefully choreographed and stage managed. The healer pretends that he knows nothing about the people he 'cures' but they have actually been selected before the show by assistants. These helpers pump the chosen ones for information soon after they arrive and this is passed on to the healer. More often than not, it is those with the least disturbing or grotesque deformities or illnesses that are selected to participate in the show. The truly disabled – the quadriplegics and those with cerebral palsy are positioned at the back of the auditorium and play no role in the dramatic, photogenic happenings on stage.

Yet the most insidious aspect of these shows and a practice that has caused many fatalities is how some very ill people manage to convince themselves they are no longer ill and injure themselves or make their condition far worse. Crucially for the preacher, these symptoms do not reappear until sometime after the show has ended and everyone has gone home.

In his book *Healing: A Doctor In Search of a Miracle*, investigator and MD, William Nolen recounted the case of Helen Sullivan, a fifty-year-old woman with metastasized cancer who was swept up by the excitement of a healing service held by the evangelist Kathryn Kuhlman. To show she had been cured, Mrs Sullivan took off the brace that had supported her back for months and walked on stage in front of thousands of hysterical supporters.

'At the service, as soon as she [Kathryn Kuhlman] said, "Someone with cancer is being cured," I knew she meant me,' Mrs Sullivan later explained. 'I could just feel this burning sensation all over my body and I was convinced the Holy Spirit was at work. I went right up on the stage and when she asked me about the brace I just took it right off, though I hadn't had it off for over four months, I had so much back pain. I was sure I was cured. That night I said a prayer of thanksgiving to the Lord and Kathryn Kuhlman and went to bed happier than I'd been in a long time. At four o'clock the next morning I woke up with a horrible pain in my back.'[2]

X-rays later showed that Mrs Sullivan had compacted a vertebra which was already weakened by cancer. She died two months later.

The charismatic or evangelical healing circus is the most dramatic and disreputable face of faith healing and to many serious well-intentioned practitioners it is itself a cancer that generates bad press for them. Healers such as Cerullo, Pat Robertson and others have legions of critics but still they thrive. After one televangelist, Oral Roberts, demanded that his followers should send him cash lest he perish, car bumper stickers began to appear brandishing the message 'LORD', standing for: 'Let Oral Roberts Die.'

Healing, in all its forms, is very much part of the 1990s ethos and, not surprisingly, it has become increasingly popular in recent years. According to the Federation of Spiritual Healers, there are more than 20 000 practitioners of some form of faith or alternative healing in Britain and a recent survey showed that 50% of people believe that spiritual or faith healing works.[3] Many celebrities subscribe to a wide-ranging selection of treatments and gurus with very good PRs. One recent newspaper reported that: 'At smart dinner parties or girly gatherings, the talk is all of toxins and unblocking, of quests and negativity. Bulging address books reveal aromatherapists and reflexologists, the best place for colonic irrigation or pendulum swinging. Tracking down the latest healer has become a kind of spiritual shopping.'[4]

Bill Clinton is said to be a follower of holistic medicine, actresses such as Demi Moore and musicians like ex-Beatle George Harrison have their favourite therapies and treatments, but in the late-1990s what was once the preserve of the rich and celebrated has filtered into middle-class circles.

There are a number of reasons why such regimens are becoming so successful. Firstly, people who either have every material benefit given to them or who have struggled to achieve the comforts of modern life still feel that something is missing – an element that will satisfy their spiritual needs. Secondly, aromatherapy, rebirthing, crystal therapy or any one of the dozens of modern variants of traditional therapies are easy. They require little effort and can be bought almost off the shelf – so much easier than working out, dieting or following the teachings of an old-style guru such as George Gurdjieff, who preached that it was 'impossible to achieve the aim without suffering.' Finally, following the teachings of a health guru absolves the individual from responsibility for their failings. Anything but total fitness and perfect health can be blamed on environmental toxins, the air, the water, stress, but never anything lacking in their own psychological make-up.

When viewed alongside the fads, crazes and more insidiously, the hundreds of cases of false cures and actual harm that many evangelical healers are responsible for, it is easy to see how the whole issue provokes extreme emotions both for and against. It is also easy to see how legitimate healers who believe they have a genuine gift (and make far less money than the big-time showmen), would feel aggrieved.

The showmen exploit simple biology and psychology to achieve their temporary wonders. The reason Mrs Sullivan was able to take off her back brace and felt no pain until much later was because her body was flooded with endorphins. As we saw in Chapter 11, these are produced by the body during times of extreme physical stress and act as pain-blockers. Experiments conducted during the early 1980s showed that the endorphin level of the subject is directly related to what has been dubbed the *placebo effect*. This describes the situation where a patient feels better simply because they believe in the treatment. A simple experiment shows how it works.

A group of patients are divided into two groups A and B. Group A are treated with a recognized drug and are told so; group B are given vitamin tablets but told that they have been given the same regimen as group A. It has been found that there is almost no difference in the response to the treatment. But if the patients in group B are told that they are only being given vitamin pills, their recovery rates are far lower.

A recent study has shown the placebo effect reduces pain because the subject releases endorphins and that these can be inhibited by a drug called naloxone.[5] Further experiments involving rats have demonstrated that this release of endorphins is a conditioned response that can be learned and controlled.[6]

If we take Mrs Sullivan as an example of how the placebo effect can work; the excitement generated by the healer and the audience triggered the release of high levels of endorphins into the blood. This, coupled with a strong desire for the healing ability to be genuine, was enough to swamp any pain she would have otherwise felt when she removed the brace and left her wheelchair. It was only several hours later when the endorphin levels had dropped sufficiently that she would have begun to feel the intense pain produced by a collapsed vertebra. Naturally, few others in the audience of thousands that night ever heard the full story and the agonies the poor woman suffered; for them, Kathryn Kuhlman was a miracle worker.

The energy produced by evangelist shows is real and undeniable, but it is not what the evangelical preachers claim it to be. After attending a performance given by Pastor John Arnott of the Toronto Blessing, a deeply sceptical and cynical journalist wrote recently that such a show was 'a profoundly unpleasant experience. I felt I had wandered inside a bizarre mass psychosis, which to resist required an astonishing degree of emotional energy. I felt drained, used up, exhausted.'[7]

There are other factors at work in these situations. Fraud and individual conviction play a major role, but timing is also important. All serious diseases follow a non-linear pattern. That is, within what may be a general decline, the patient suffers bad periods and good periods. Even those suffering fatal illnesses and who die from the affliction eventually, go through times of better or worse health. Logically, an individual is most likely to feel the need to visit a faith healer when they are at a particularly low point. Often faith healers are used as a last resort, when a patient feels they cannot possibly get any worse. If an individual then attends an evangelical performance or visits a private healer, the chances are they will soon enter a less severe spell of illness purely as a consequence of the natural waxing and waning of the disease, (or else they will die). But, not realizing that this is a natural pattern within the course of the disease, the patient will feel justified in attributing the sudden improvement to the healer, and the healer will of course be quite happy to take the credit.

A different kind of healing is that generated by a belief in the holiness or the curative powers of a particular location. Usually these special places are identified with mythical or religious events and hold a powerful attraction for the desperate believer. Such places play the same role as the charismatic healer; they trick the faithful into believing they are cured because they acquire a short-term pain inhibition from the excitement of being there or by taking part in a ritual. This effect is particularly powerful when the healing site has strong religious connotations.

The most famous example of a Christian shrine is Lourdes in France. Lourdes became a centre for pilgrims after a girl claimed a visitation of the Virgin Mary there in 1858. Within twenty years a shrine had been constructed on the site and today it has over five million visitors a year. Although tens of thousands claim to have experienced 'miracles' at Lourdes each year, there have only been 64 cases accepted by the Catholic church over the past 130 years. These have been passed and accepted as miracles by a specially appointed medical group which was formed in 1947 called the Medical Bureau. To qualify as a miracle, an incident must conform to a set of tough criteria established by Cardinal Lambertini (later Pope Benedict XIV) in 1758. These state that the disease must be incurable and unresponsive to treatment, it must be at an advanced stage but not at the point where it has resolved itself and any medication the patient has received must be shown to have already failed. Finally, the cure must be instantaneous and total.

Even the 64 approved cases are open to criticism and some claim that as medical knowledge has improved, the cases passed in the early days of the Bureau would not now be allowed. It is also telling that there has never been a clear-cut, obvious miracle that was undeniable and without controversy. As the writer Anatole France commented when visiting Lourdes at the end of the nineteenth century and seeing the discarded crutches of the 'cured', 'What, what, no wooden legs?'

Lourdes and other shrines energize the wishes of the desperate; they focus the needs and desires of the very sick and can, for a short time perhaps, create the outward signs of a cure – hence the discarded crutches. But, just as with the televangelist con-show, follow-up investigations into what later happens to pilgrims are rare.

One investigator, a doctor from Southampton, Peter May, has spent twenty years taking the trouble to carry out detailed investigations of those who claimed to have been cured at Lourdes. He has come to the conclusion that there is no substance to big-time faith healing, to the televangelists and the world-renowned shrines, and sees the entire phenomenon as a mish-mash of misunderstandings and misinterpretation.

'The fact is that a large proportion of medical conditions get better anyway, treated or not,' he said in a recent magazine article. 'I've seen supposed cures of sarcoidosis [a disorder of the lymph nodes] describing it as a "rare and potential killer disease" – when we know that 80% of cases get better, with no treatment. Migraines, backaches, nausea, phobias and eczema regularly feature on lists of cures, but they all come and go, apparently independently of treatment. In usually irreversible conditions, such as sensorineural deafness, for which healing cures have been claimed, there are cases of spontaneous remission in medical literature.'[8]

So, if this is the case for large-scale phenomena such as Lourdes and the preacher-healers who are supported by millions around the world, what of the individual healer working alone and dealing with individual patients? Is there any chance that some individuals are truly gifted, or that there are special techniques which science cannot explain that do facilitate genuine cures?

Almost all evangelical healers claim their spurious gifts come from God, that they are mere conduits who channel the energy given to them, but there are also a number of well-intentioned, non-evangelical healers who believe they possess this power.

One of the most famous is Matthew Manning, who as a schoolboy started to become the centre of apparent poltergeist activity at his parent's home and at his boarding school. He soon developed an ability for automatic writing and for reproducing works of art in the style of the great masters. Manning has no inherent artistic ability, but the paintings he reproduced were of the highest quality and in a range of styles.

Today he is a professional healer and claims that he channels the energy he used for automatic writing and painting into curing the ills of others. 'It is not me,' he has said. 'I simply switch on the energy.'[9]

Although his abilities are open to the same questions and doubts that confront all faith healers, his inexplicable artistic skills and his laboratory-tested psychokinetic demonstrations seem to show that something incredible (but not necessarily supernatural) is happening within his mind.

The problem with the notion that Matthew Manning or any other claimant as a genuine conduit for a supernatural force is the question of relevance. Quite simply: Why should a spirit, a God or an alien entity want to assist in these processes? Although science strains to explain how ghosts could be play-backs of past events or how poltergeist activity is probably collective hysteria, the acceptance of immortal spirits, anthropocentric gods or concerned aliens is simply illogical and based entirely upon an exaggerated sense of self-importance. Even if for a moment we accept that there may be a divine being, even if we allow ourselves to believe in life after death, why should an all-powerful universal being care about the fate of a single organism on Earth?

The Earth is a tiny planet, one amongst perhaps millions of inhabited worlds in a near-infinite universe. Does it make any sense to suppose that God would be interested in the chilblains of Mrs Smith at Number 46?

Yet Matthew Manning has demonstrated amazing powers as an artist and now claims to be applying the same gift to a different end. It is perhaps significant that many sceptics refuse to even mention him, knowing they are unable to dismiss him using present-day scientific understanding. Is ability like Manning's truly a gift from the gods as some claim, or is it rather an example of the enormous potential of the human brain? We use little more than 10% of our brain capacity: could there be hidden regions where exceptional powers of memory, of visualization, and of dexterity would allow an ordinary individual to reproduce a Leonardo perfectly? And could similar abilities explain the claims of faith healers?

It is because of this question that those healers who believe they are able to focus energies within themselves to cure the ills of others, are far more convincing than the evangelical healers. Yet interestingly, those who practise healing often have no idea where the energy they use comes from.

In many cultures there is a long tradition of employing 'internal healing energies'. The shaman of African tribes have practised healing techniques for thousands of years and the alternative medical practices of acupuncture, crystal therapy and systems such as shiatsu all rely upon balancing the internal 'energy fields' of the body.

But what are these energy fields? Orthodox medicine has failed totally to locate them, to detect or observe them, yet techniques such as acupuncture are being absorbed into the body of conventional medicine because they have proven so successful.* Is acupuncture really another example of a placebo, because, if we accept that acupuncture works (which is, after all, a technique dependent upon the notion of an undetectable energy system), then where does science and medicine draw the line? If acupuncture, why not crystal therapy which suggests that certain minerals can be made to resonate with the 'human energy field'? And, if crystal therapy is acceptable, what should we make of healing by the laying on of hands?

The closest we have come to any form of observing an energy field around animate objects is through the technique of Kirlian photography.

The method of photographing what enthusiasts believe to be a representation of an aura surrounding living things is credited to a Russian couple, Semyon and Valentina Kirlian, who produced their first image in 1939. The Kirlian photograph is created by passing a small electric current through living tissue which produces a brightly coloured pattern that appears to be affected by the emotional condition and the general health of the subject.

Enthusiasts suggest that this image represents the energy field of the person, animal or plant subjected to the electric current. Their originator, Semyon Kirlian, said of the images: 'The inner life activities of the human being are written in these light hieroglyphs.'[11] The sceptics, however, have another explanation and claim the excitement Kirlian photographs have generated should be

* According to a report in *The Mail On Sunday* in December 1991, there are upwards of forty NHS doctors who are also faith healers.[10]

seen as nothing more than a simple misinterpretation of the scientific facts.

The image is created by the electric current ionizing the air surrounding the object and the shape is determined by its outline. Scientists believe the bright colours and the often dramatic images are greatly enhanced by the presence of moisture. This, they say, is why living objects create strong Kirlian images but inanimate objects do not. They further suggest that the reason the image appears to be influenced by mood is due to the fact that moisture levels close to the skin are subject to emotional changes – clammy hands being a sign of anxiety for example. Further doubt comes from the fact that if the Kirlian photograph is taken in a vacuum no image is visible, which implies the brightly coloured fringe is an effect produced entirely by the ionization of particles in the air and nothing to do with the living object itself.

Supporters of the aura theory have countered this with an impressive body of research conducted since the 1930s which has revealed a growing collection of odd effects. The most striking result, and one which the sceptics have so far failed to explain fully, came from an experiment first carried out on a leaf. The leaf was photographed using the Kirlian technique and showed the usual glowing outline. Then a section of the leaf was cut away and the remaining section re-photographed. To the astonishment of the experimenters, the Kirlian image mirrored the full outline of the uncut leaf as if the missing part was still attached.

This and other experiments have led researchers of the technique to suggest the existence of what they have dubbed a *bioplasmic field* and the notion of *bio-energy* – new names for the traditional ideas of the 'energy field of the body' and the energy that many healers believe to be responsible for their abilities.

As we saw in Chapter 7, the mind seems capable of controlling the body to a surprising degree and this ability can be honed and developed in appropriate individuals. In the same way, it may be possible that a talented highly trained healer could instruct the body to repair itself or defend itself against disease by suggestion – amplifying the wishes of the sick individual. This may take the form of hypnosis, in which the patient and the healer each play a role in creating a barrier against anything from an allergy to a cancerous tumour.

In one sense this is not a supernatural phenomenon but one utilizing extreme powers of the mind and body and exploiting metabolic self-control. It might seem supernatural because it is far from commonplace, but it does not require occult agents or even religious observance or devotion.

During the 1950s, the psychiatrist Stephen Black conducted an impressive series of experiments to demonstrate the power of hypnosis in healing and disease prevention. He took a group of subjects who were susceptible to hypnotic suggestion and also suffered common allergies. In traditional fashion, he divided his subjects into two groups. The first were hypnotized and the material to which they were allergic was injected just beneath the skin. This allergen very quickly produced a characteristic red 'wheal and flare' which increased in intensity over a period of twenty minutes.

Black then repeated the test with the second group but induced hypnosis with the suggestion that they would not react to the allergen when it was introduced. The material was injected under the skin, and none of the hypnotized subjects showed any sign of an allergic response. Black then conducted biopsies on both sets of subjects and discovered that in the second group, the chemicals responsible for the lesions had been blocked.[12]

The placebo effect upon which most sceptics rely could not have played a role in this set of experiments because the hypnotized subjects could not have been aware of the process. However, the fact that they were told they would not respond to the allergen could have been the trigger for the release of the all-important endorphins that seem to be the principal agent behind the placebo effect.

The growing field of psychoneuroimmunology (PNI) is beginning to throw up new and exciting variants on this theme. PNI attempts to explain how mood and emotions can influence the body's defences and responses to disease. Until recently it was believed that the body defended itself against viruses and bacteria independently of the brain – that specific regions under attack responded autonomously by triggering the appropriate biochemical agents. Now, neurologists are changing their minds and coming to accept a role for peripheral nerves that link the thymus gland, the lymph nodes and bone marrow which produces white blood cells (lymphocytes) – the body's army, navy and air force.

During times of stress or depression, the immune system is suppressed by a group of biochemicals called *corticosteroid hormones*; by suppressing the action of the lymphocytes these chemicals weaken our natural defences. Conversely, at times of extreme happiness or euphoria we produce a chemical called *interferon*, (which is now gaining use by doctors as a major cancer-suppressing material) and assists the immune system to defend against attack.

The release of these chemicals can be triggered by a variety of sources. If we consider only the 'positive' biochemicals (those that enhance health), endorphins are released in a rush over a short time frame by transitory excitement – a response to strenuous physical activity or the buzz created at an evangelical meeting. These chemicals get us through emergencies and give us an extra boost and it is because of this that the televangelist is a danger, as the unfortunate Mrs Sullivan discovered.

More useful in the long term are those agents released slowly and over longer time periods. These could be encouraged by a caring healer, by a belief in a technique such as acupuncture, aromatherapy or colonic irrigation. Such techniques provide a more controlled, balanced and more easily assimilated neurological response. Surprisingly perhaps, pets have been shown to be another source of comfort for the ill; they relieve stress and offer what one psychologist calls, 'a form of unconditional love that assists the immune system.'

So it would seem that almost all aspects of faith healing may either be dismissed as dangerous exploitation or explained by modern neurophysiology. The placebo effect is a term that is over used and often inappropriately employed, but there is a growing body of evidence to show that the brain does play a very active role in bodily health, and with training, this ability could be gainfully exploited. However, there remains one form of faith healing that has so far defied scientific explanation and has become another topic conveniently ignored by sceptics.

One day in 1993, racehorse breeder Jan Piper discovered a lump over the eye of her favourite breeding mare, Jessica. She was immediately concerned and feared the worst. The vet arrived later that morning and soon found several more lumps in the horse's throat and groin. A week later, Jan Piper had the diagnosis. The horse was suffering from *equine lymphosarcoma*, a rare form of cancer which affects the white blood cells before attacking the internal organs. It is fatal and quite incurable.

In despair, and on the eve of Jessica being put down, the breeder heard about Charles Siddle, an animal healer. Despite the scepticism of her husband, Jan Piper called Mr Siddle and the next evening, he arrived at the stable. He was there for less than half an hour. After placing a large crystal on the floor of the stall, he laid his hands on Jessica and ran his palms over the horse's coat, along her mane and across her eyes. 'She'll probably be tired tomorrow,' he said quietly, 'but after that she'll be fine.' Then he left.

Next morning, Jessica started eating for the first time in weeks and within days she was completely cured. She now has a foal and is still in the best of health.

Animal faith healing is a true mystery. Because it involves animals and not free-thinking humans, sceptics cannot accuse believers of wishful thinking or even the placebo effect. Some doctors suggest that apparent miracle cures such as Jessica's are an example of spontaneous remission, and indeed, cases of sudden natural cures in humans and other animals are well documented, if rare.

The British Veterinary Association refuses to acknowledge the work of animal faith healers, but the prestigious Royal College of Veterinary Surgery has recently allowed them to work with animals in the presence of a qualified vet. But frustratingly for the scientific investigator, the healers themselves talk in the same vague terms as those who deal with human patients. Like them, they seem to have little idea how they do what they do. 'I don't really know what happens when I heal,' one said during a recent television documentary on the subject. 'The energy comes through my hands, so I guess I act like a channel.'

It has been suggested that the healing process for any animal is the same – a matter of 'realigning the aura' or readjusting the energy field. The problem with this attempt at an explanation is that enthusiasts are again trying to solve a mystery using a phenomenon that itself has no scientific support. It is as bad as suggesting that ghosts are definitely the spirits of the dead – faith healing is a mystery and so is the concept of the energy field or the aura.

Yet Charles Siddle continues to have remarkable success with the animals he treats. One prize racehorse he attended had a ligament so badly damaged the horse could hardly walk and the lower leg was severely swollen. After one visit, the swelling had almost gone and the horse was cantering around the paddock as if it had never been injured. This certainly cannot be explained by spontaneous natural remission.

There is other evidence to suggest the placebo effect is not the only mechanism by which an individual can control symptoms or keep disease at bay. In Canada, a healer called Oskar Estebany and a researcher, Dr Bernard Grad working at McGill University in Montreal have found that mice are susceptible to what appears to be Estebany's healing powers. Three groups of 16 mice each had a small patch of skin removed. The first group were subjected to Estebany's healing touch, the second group were warmed (to rule out the possibility that it was heat from the healer's hand that affected them) and the third group were ignored. The first group showed significantly faster recovery rates than the other two.

In a similar experiment, this time using human guineapigs, a team based at the JFK University of California appear to refute completely the idea that all unorthodox healing is simply placebo. A piece of skin was removed from the shoulders of 44 subjects who were told they were involved in an orthodox medical experiment. Each day the subjects were asked to place their wound in a window in the belief they were being photographed. Some of the volunteers were actually being treated by a healer, others were not. The wounds of those singled out for psychic treatment recovered far faster than those who were not.

Some scientists have taken the investigation of healing powers even further and conducted experiments on plants. Dr Tony Scofield and Dr David Hodges working at London University collaborated with a healer named Geoff Boltwood. The experimenters took a batch of watercress seeds and treated them with salt water to 'make them sick'. Then, following the usual pattern, the sample was divided into two groups. Boltwood then held half the seeds in each hand. The first group he tried to heal by concentrating his abilities and directing his energies towards them; the second group he held but ignored. After six repeats of the experiment conducted under scrupulously controlled conditions, the team were surprised to discover that in five of the tests the seeds subjected to Boltwood's efforts grew at almost double the rate of the others.

It could be that the researchers missed some crucial element in these experiments and their results have been vilified in the scientific press, with some sceptics just stopping short of denouncing them as frauds. But theirs is not an isolated experiment; teams in the United States and Europe have shown similarly exciting results.

How any of these non-placebo effects could work remains a mystery. Enthusiasts of the phenomenon of psychic healing see them as clear evidence that the healer is either able to channel external energies or is in some way capable of using their own energy fields to correct those of sick people, animals or even plants. The sceptic points to the valid fact that these are tiny trials, and despite the best efforts of the experimenters, errors could have been made or unknown extraneous, but quite natural factors could lie at the root of the effect.

Like telepathy, precognition and many of the other topics covered in this book, healing is another borderline activity. And so it seems to be with almost all aspects of the paranormal. Alien vehicles have not landed in Parliament Square, the sick do not grow severed limbs spontaneously and no one, not even self-publicists like Uri Geller, has managed to give a clear undeniable demonstration of paranormal abilities. This is very disappointing but it should not lead us directly to the conclusion that all occult matters are bunkum. Much of it certainly is, but not all. As the astronomer Sir Martin Rees has said: 'Absence of evidence is not evidence of absence.'

References

1. *The Week-End Book*, 1925, p. 158.
2. William Nolen, *Healing: A Doctor in Search of a Miracle*, Random House, New York, 1974, pp. 98–99.

3. Jerome Burne, 'Do You Believe?', *FOCUS*, December 1993, pp. 36–41.
4. Cassandra Jardine, 'Have Faith – and Ease Those Ills', *The Daily Telegraph*, Friday 31 May 1996, p. 20.
5. R. Gracely et al., 'Placebo and Naloxone Can Alter Post-surgical Pain by Separate Mechanism', *Nature*, 306, pp. 264–265.
6. L. Watkins and D. Mayer, 'Organization of Endogenous Opiate and Non-opiate Pain Control Systems', *Science*, 216, pp. 1185–1192.
7. John Sweeney, 'Gigglers For God', *Life Magazine (The Observer)*, 3 March 1996, pp. 30–33.
8. As ref.3.
9. Matthew Manning, *In the Minds of Millions*, Allen, London, 1977.
10. As quoted in John and Anne Spencer, *The Encyclopaedia of the World's Greatest Unsolved Mysteries*, Headline, 1995.
11. S. Ostrander and L. Schroeder, *Psychic Discoveries Behind the Iron Curtain*, Englewood Cliffs, New Jersey, Prentice-Hall, 1971.
12. A.A. Mason and S. Black., 'Allergic Skin Responses Abolished under Treatment of Asthma and Hayfever by Hypnosis', *Lancet*, i, 1129–1135, 1958.

Afterword

The success of the *X-Files* is a phenomenon in itself. In Britain, the second series achieved viewing figures of more than 7 million on the minority channel, BBC2; in Japan, 120 000 videos were sold before the programme was even broadcast and in Spain, it is the most popular programme *ever*. What is the secret?

As well as having a sense of humour, charismatic actors and well-developed characterization, the *X-Files* combines three aspects that touch the public consciousness during the late 1990s. Firstly it taps into the notion of conspiracy theories/FBI machinations and connections with fundamentalist militia and underground terrorist intrigue. Secondly, it is linked heavily with the Internet and has appeared on our screens just as this embryonic technology is emerging into something tangible. Lastly, and most importantly, it deals with the paranormal and the outer rim where science meets science fiction. It is this which has really sold the programme and catapulted it into the mainstream.

It seems that around every twenty years the paranormal crawls from the woodwork and comes blinking into the full glare of publicity. The 1930s was a big time for the supernatural with such figures as Aleister Crowley then at his peak. The 1950s saw the first blossom of the UFO myth and the earliest tales of close encounters as popularized by George Adamski. In the early 1970s the supernatural surfaced again and amalgamated at times with science fiction, giving us a succession of popular novels, films and TV series – *The Exorcist*, *Carrie*, *UFO*, to name but a few. Now, it has all come around again in a far more sophisticated form with some new elements thrown in that could not have been imagined in previous incarnations – genetic mutations, the Internet, cybertechnology.

So, why does the subject of the paranormal intrigue and captivate us so? Why do we tune in to the *X-Files*, buy the novels, wear the T-shirts?

It is partly a generational thing. The fact that the paranormal reappears in the public imagination about once every two decades is no coincidence and for many young people, all aspects of the investigations of Mulder and Scully are totally new and original. But it runs deeper than that. In each of us there is a yearning for something larger than life, something beyond the mundane.

Perhaps, as life becomes more comfortable, we need to find something extra, something beyond ourselves. Most of us cannot find this extra element in our 'real' lives so we look for it elsewhere – we escape.

Like most people, I really wish many of the ideas of the occultists were actual. It would be such fun if ghosts existed (other than as hypothetical played-back images). It would be tremendously exciting if aliens really were here and we could communicate with them. How much more entertaining life would be if we could develop big-scale, usable telepathic powers. I'm not so keen on spontaneous human combustion, but precognition could be handy if used properly and time travel would be a dream come true. Sadly, for most of these, the evidence points the other way. There may be some effects on a small scale with phenomena such as telepathy and PK. We certainly can control our own bodies to a degree and some gifted individuals may be able to enhance this ability in others. There are almost certainly alien civilizations not far from Earth (in cosmic terms) and they may well have visited, but there is no huge conspiracy or cover-up hiding a silent invasion. In the final analysis, *Independence Day*, *Alien Nation* or *War of the Worlds* are fantastic entertainment, but once you leave the cinema, close the book or switch off the TV, they have no existence beyond memory.

Yet, none of this stops us striving and I am not one of those trained as a scientist who totally dismisses the paranormal, (as I hope you will have already gleaned). It is important we all keep open minds, but also fully functioning ones. There is something off-putting about the smugness of a small minority of New Age practitioners, a smugness borne of ignorance and dysfunctional mentality that gets none of us anywhere.

By all means, keep questioning, prodding and investigating. Keep watching the skies – if for nothing else but a window onto the eternal.

The author

After a disreputable youth spent touring and recording with the Thompson Twins pop group, in 1982 Michael White graduated from King's College, University of London. He then moved from a musical career to lecturing chemistry at d'Overbroeck's College, Oxford, where he eventually became Director of Scientific Studies in 1989. He began writing professionally in 1987 and is the co-author of *Stephen Hawking – A Life In Science* (1992) and *Einstein – A Life In Science* (1993). The first of these was a No. 1 bestseller and remained in the *Sunday Times* bestseller list for four months. Both books have now been translated into 25 languages world-wide and *Hawking* has sold an estimated 250,000 copies.

1994 saw the publication of *Asimov – The Unauthorized Life* (Orion) and in 1995 two further books were published *Darwin – A Life In Science* (Simon and Schuster – UK, NAL/Dutton – US) and *Breakthrough – The Hunt for the Breast Cancer Gene* (Macmillan – UK, Wiley – US). Both of these were published in the United States at the beginning of 1996.

Michael White is the science editor of *GQ*, and a regular contributor to *The Daily Telegraph*, *The Mail on Sunday*, and *The Sunday Times*. He is currently working on an alternative biography of Newton called *The Last Sorcerer* (to be published by Fourth Estate in Britain and Addison Wesley in the US) and a science thriller entitled *The Face of God*. He lives in London with his wife.